How Slow

D0538132

How Slow Can You Waterski?

Edited by Simon Rogers

guardianbooks

Published by Arrow 2006

2 4 6 8 10 9 7 5 3

Copyright © The Guardian 2006

The Guardian is a registered trademark of the Guardian Media Group Plc.
Guardian Books is an imprint of Guardian Newspapers Ltd.

The moral right of the authors has been asserted

First published in Great Britain in 2006 by Arrow Books
Random House, 20 Vauxhall Bridge Road,
London SW1V 2SA

www.randomhouse.co.uk

Addresses for companies within The Random House Group Limited can be
found at: www.randomhouse.co.uk/offices.htm

The Random House Group Limited Reg. No. 954009

A CIP catalogue record for this book
is available from the British Library

ISBN 9780099505266 (from Jan 2007)
ISBN 0 09 950526 6

The Random House Group Limited makes every effort to ensure that the
papers used in its books are made from trees that have been legally sourced
from well-managed and credibly certified forests. Our paper procurement
policy can be found at: www.randomhouse.co.uk/paper.htm

Typeset by SX Composing DTP, Rayleigh, Essex
Printed and bound in Great Britain by
Bookmarque Ltd, Croydon, Surrey

Contents

Introduction

CONTROLLED EXPLOSION PACKAGE PROVES HARMLESS
CHOCOLATE IS THE NEW HEALTH FOOD
ASTRONOMERS ARGUE OVER NEW PLANET

Do these headlines seem familiar? Have you ever stopped to think about what they really mean? Or do you just glance at the story and turn the page of the paper? Have you ever wanted to find out exactly how you control an explosion? If chocolate is really good for you? Or even what makes a planet a planet? We did, and this is how we came to start up *This Week – the science behind the news*. It began as part of the *Guardian*'s *Life* supplement, launched by Emily Wilson and carried on by me, and swiftly became one of its most popular columns, surviving *Life* to become part of the paper's main news section. And the reason for its success? It answers those nagging questions that lurk behind every news story but rarely make it into the paper.

At one level it is serious stuff. Politicians may claim that the only solution to the energy crisis is to go nuclear, but what exactly are the risks, and can we be sure that new nuclear power stations will deal with them? Every few months there seems to be new superbug set to cause a pandemic, but what is the real threat, and can we protect ourselves if we need to? At the other end, when a royal is forced to take her dog to a pet psychologist, don't you want to know what they actually do there?

Despite the scientific community's obsession with communication, there is a lot of bad, sensationalist science reporting in the UK media. 'Miracle cures' and 'startling new developments' are miraculously and startlingly frequent. You need a bit of background knowledge to untangle the ends of a story, and *This Week* is part of the attempt to provide that.

Yet all of the articles have their own existence, independent of the story which inspired them. At the time the pieces were commissioned and written, we treated them like news stories, and although they may seem short, enormous amounts of real reporting by the *Guardian*'s science team went into them every week.

We agonised over the currency of the pieces and the issues – would they stand up or just seem dated by the time the *Life* section came out? 'Do books improve your mind?' was written about a celebrity who had never read a book but was writing an autobiography. 'Can acupuncture help you to beat cocaine addiction?' was written as model Kate Moss struggled with the drug.

But current as the pieces were designed to be, most of them live beyond the week they were published. I will always be fascinated by what happens if you drill a hole in your head or how many vaccinations a baby can have.

These are the big questions of life. And the little ones too.

Simon Rogers

All the articles here were written by the *Guardian*'s science team – Tim Radford, Ian Sample, David Adam, Alok Jha and James Randerson – with guest appearances by other *Guardian* and science writers, particularly Kate Ravilious, Ben Goldacre, Sarah Boseley, Steven Morris, Lucy Rogers, Bill Hanage and Laura Bach.

Minds & Bodies

Do books improve your mind?

We all learn to read, but what happens in adult life when we fail to keep it up? Does the brain shrink like a withered prune? Studies in America found that continued intellectual activity between the ages of 20 and 60 may protect against dementia in later life. One found that continuing intellectual pursuits reduced the risk of Alzheimer's disease by a third. In another study, relatively inactive patients were 250% more likely to develop Alzheimer's.

Damaged brains can adapt and learn. Researchers who have used brain scanners have found that other parts of the brain can compensate. But exercising the brain, in much the same way as one would exercise a damaged muscle, perhaps by repeating a list of items, does not help regrowth.

Are you going to benefit more by reading Shakespeare than *Vogue*? It probably doesn't matter as long as the brain is exposed to new information that stimulates your cells.

Luckily, physical activity also counts. Whether physical exercise is as beneficial as intellectual activity remains unknown.

Do animals make you feel better?

The idea might sound like new age mumbo-jumbo. But scientists now believe that swimming with dolphins really does alleviate depression.

It supports a theory put forward by the sociobiologist Edward O. Wilson. According to his idea of biophilia, human health and well-being are dependent on our relationships with the natural environment. This means that animals and natural scenery help us feel better, and our happiness around nature is somehow hard-wired into the brain. A growing body of clinical evidence suggests that Professor Wilson might have a point. In a paper published in the *American Journal of Preventive Medicine* in 2001, public health scientist Howard Frumkin of Emory University, Atlanta, reviewed the evidence for the health benefits of four kinds of contact with the natural environment: contact with animals, plants and wilderness and viewing landscapes.

He pointed to research which concluded pet owners have fewer health problems than non-pet owners. They had, for example, lower blood pressure, improved survival after heart attacks and better ability to cope with life stresses. At Purdue University in Indiana, patients waiting for dental surgery were found to experience a clinically significant drop in blood pressure after staring at fish in an aquarium for 20 minutes. In another study, University of Washington scientists found that children with autism who were allowed to play with dogs became more verbal and engaged with therapists.

In Japan, researchers compared the responses of people who looked at a hedge with those staring at a concrete fence. The former experience caused relaxation, while the latter produced stress. Similar responses occurred when subjects looked at a vase filled with flowers as opposed to an empty pot.

Why any of this should happen is largely unknown but Professor Frumkin had some ideas. 'Early humans found that places with open views offered better opportunities to find food and avoid predators,' he said. 'But they needed water to survive and attract prey, and groups of trees for protection. Modern research has shown that people today, given the choice, prefer landscapes that look like this scenario.'

Can you die from heartbreak?

With the caveat that it is difficult to establish a link between emotional stress and physiological health, all the evidence suggests that the answer is yes.

The first study to look at the issue was published in the *British Medical Journal* in 1969. Researchers followed 4,500 widowers, all 55 years or older, for nine years and found that the risk of dying in the first six months after bereavement was 40% higher than expected, then it gradually fell back to normal.

A bigger study, published in 1996, confirmed these results. Scientists looked at more than 1.5 million people aged between 35 and 84, and found that, in the six months after losing a spouse, the risk of dying from a heart attack increased by 20 to 35%. They also found that the risk of dying from an accident, violence or from alcohol-related problems nearly doubled. And in most cases, the risk of death was greater for men.

Why bereavement might trigger death or illness is largely unknown, but speculations are rife. When people lose the lifetime support offered by a partner, they are more likely to get stressed. This might have acute effects on the body and, the more elderly the person, the more pronounced those effects may be.

People suffering from stress due to losing a loved one have reported a range of health problems – from gastro-intestinal complaints to muscular pains. The sudden stress could also trigger more serious underlying problems, such as heart disease.

How psychological pain turns into a physical problem is also an active area of research. The accepted wisdom is that the brain, after registering the psychological and social variables around it, will signal instructions to release certain hormones into the bloodstream and these affect mood as well as subsequent health.

Psychologists have found, for example, that people going through a rough patch in their relationship were more likely to catch a cold or flu. In a study of 2,000 people in various emotional states at the Medical Research Council's social and public health sciences unit in Glasgow, researchers found that stress or bereavement was linked to a decrease in the levels of an antibody called immunoglobulin A, which is the body's first defence against foreign microbes.

Why this happens is unknown, but researchers believe it might be down to high levels of the hormone cortisol, which tends to increase during stressful situations.

Does having wonky elbows matter?

That depends. Are you a man? Do you have a wife or girl-friend? And, most importantly, are your ears and fingers as mismatched as your arms?

If the answers to all of the above are yes then your (unbalanced) ears will have pricked up at the news that your partner is most likely to be unfaithful. A study of 54 couples by the University of New Mexico found that women whose partners have mismatching ears, fingers or elbows tend to fantasise about sex with other men when they are ovulating. Those whose men happen to be neatly proportioned do not, and still prefer their partners to other men, even in the middle of their monthly cycle.

Studies of sexual desire are not new. Dave Perrett at St Andrews University suggests that women prefer symmetrical faces because this indicates healthy genes in their partner.

Sex hormones are linked to feminine and masculine facial features – youth and fertility signalling good long-term health. By exaggerating such facial features, researchers have found that women are attracted to strong masculine faces but too masculine a face can be a turn-off, indicating a cold and dishonest mate.

Can you stop yourself sweating?

If horses sweat, men perspire and ladies glow, then all three have their autonomous nervous system to thank. That means that sweating (or perspiring or glowing) is a reflex action and independent of direct messages from the brain. Some people have a more responsive nervous system than others, so while some are cool under pressure, others may find embarrassing stains on their shirts. And alcohol can effectively reset the nervous system to produce yet more sweat.

But for politicians caught sweating on prime-time news broadcasts, short of crash diets, lowering the lights and asking the audience to leave, is there anything that can reduce the visible proof that politics is 1% inspiration and 99% perspiration?

'There are a couple of medications that might work,' says Antranik Benohanian, a dermatologist at Montreal University Hospital, who has treated more than 5,000 patients with hyperhydrosis, the clinical term for excessive sweating. Some of these can be used on specific areas of the body, mainly by targeting a neurotransmitter called acetylcholine, which is produced by nerve endings under the skin and turns on the taps when it reaches the sweat glands. Applying it to the hairline the night before a big speech could prevent a sweaty forehead the next day. 'But there is no solution without side effects,' warns Benohanian. Some treatments merely shift the damp patches to other areas, and some induce blurred vision and a dry mouth – hardly inspiring stuff for a would-be prime minister.

Another possibility is the botox injections favoured by the wrinkle-free rich. The toxin knocks out acetylcholine transmission in the target area, offering up to a year of reduced sweating. Liposuction can also destroy nerve endings beneath the skin, stopping the sweat message from being sent.

Can a blow to the chest stop your heart?

'It requires a lot of force in one place on the left-hand side of the chest,' says John Martin, a cardiologist at University College London. 'It's very rare.' Unfortunately, the odds worked against a young cricketer in Liverpool, who was hit in the chest by a ball. He died after his heart stopped beating.

'One in a million cricket balls hitting you on the chest would have this effect,' says Martin. 'Each cardiologist would see one in a career.'

The heart beats because of an electrical impulse generated at the top of the organ in the atrium. This electrical signal passes down the atrium and then into the ventricle, essentially a pump made of muscle. The signal ensures that the heart contracts all at once to force blood out into the bloodstream.

Under certain conditions, the signal is disrupted, most commonly through disease but, very rarely, through an external stimulus.

'The impact of the ball has caused disorganisation of the electrical signal passing through the heart,' says Martin. 'Each little muscle fibre contracts independently of all the others. So there's a great fluttering of this great muscle instead of a contraction.' This flutter, or ventricular fibrillation, is the most common cause of death in the hours after a heart attack.

'The tragedy is that it can be reversed fairly easily by a defibrillator,' says Martin. Immediately after an accident, keeping the heart pumping until medical attention arrives can save lives. Even if a heart's electrical activity is disrupted, pumping the patient's chest can keep blood flowing to the brain until medics arrive with a defibrillator. This device works by shocking the heart into re-organising its electrical activity.

'Everybody should learn how to do cardiac resuscitation, how to go to a young man like this with no pulse, to press rhythmically on his sternum.' says Martin.

How long can you survive in a freezer?

A question that Richard Carter must have asked himself when some kids locked him in his ice-cream freezer.

Carter, who was trapped in the −28°C chamber for 15 minutes, told a newspaper: 'Another 15 minutes and I'd have been a goner.'

The first sign of trouble is frost-bite, says Bill Keatinge, physiologist at Queen Mary, University of London. In extreme cold, our bodies shut down the blood supply to our skin, and because our fingers are so small, they can freeze quickly if not covered up.

'In experiments, I've frozen my little finger repeatedly, and it only takes about 70 to 80 seconds,' says Keatinge.

Frozen fingers are a big issue in Yakutsk in eastern Siberia, the coldest town in the world. Drunks who collapse outside often have frozen fingers by the time they are found. 'The local doctors do between one and three finger amputations a day, and it's a small town,' says Keatinge. 'It's a problem all over Russia.'

While shivering keeps you warm, boosting your body's heat production tenfold, it uses a lot of energy, so can be exhausting.

When shivering stops, it's time to worry. Even if you are fat, you will begin to lose heat quickly, falling into a state of hypothermia once your core body temperature drops below 35°C.

As the body cools further, breathing becomes laboured and it becomes hard to think straight. Ultimately, the heart muscles begin to seize up, and because blood is then pumped around the body so inefficiently, tissues and organs fail through lack of oxygen. 'You'd be in real trouble within hours at −28°C,' said Keatinge. 'I'd be amazed if anyone survived as long as a day at that temperature.'

How long can someone survive without water?

Not as long as aspiring Buddhas may claim. Reports from Nepal told of a teenage boy meditating for the last six months and said to have not drunk any water for the entire period. Suspicious locals asked for a scientific examination to determine if the boy was managing without water.

The magician David Blaine survived 44 days without food, losing one quarter of his body weight, but keeping a healthy body mass index. In 1976 obese people were put on an experimental starvation diet, with absolutely no food, for 40 days, and none of them had any trouble surviving. 'It is possible to last much longer without eating than without drinking,' says Martha Stipanuk, from the division of nutritional sciences at Cornell University in New York. But it does depend on your initial body condition. 'A weak elderly person or thin young person might not be able to go very long without food,' she adds.

The problem for wannabe Buddhas is that surviving for weeks without water is not an option. 'People can last a few days without water depending on the environment in which they find themselves and whether [they are] injured or not,' says Jeremy Powell-Tuck, professor of clinical nutrition at Barts and the London Queen Mary school of medicine, who supervised Blaine's recovery.

Someone sitting quietly under a shady tree will be better off than an explorer caught out in the middle of a blazing desert, but none the less they won't be able to survive for six months without a sip of water.

'Without water anyone will run into problems pretty quickly. Their blood volume will shrink and their water and electrolyte balance will be upset. Eventually the body will just go into shock,' says Professor Stipanuk.

How tall can a human grow?

History provides a few pointers. According to the Bible, the tallest man was Goliath at 'six cubits and a span', which, depending on whose conversion you believe, puts him somewhere between nine and a half and eleven feet tall. Sadly though, the Bible was not peer-reviewed, so Goliath must be disqualified.

The tallest man on record is Robert Wadlow, an Illinois man who died at 2.71 m (8 ft 11 in) in 1940 at the age of 22. The record may not stand for much longer, however. Leonid Stadnyk, a 33-year-old living in a remote village in Ukraine, hit the news as the world's tallest living man. At 2.54 m (8 ft 4 in), he is just 17 cm short of Wadlow's record. In the past two years, he has grown 30 cm.

Like Wadlow, Stadnyk owes his extraordinary height to a tumour on his pituitary gland. The tumour churns out growth hormone but it's a secondary effect that leads to the runaway growth that doctors call acromegalic gigantism.

Normally, the growth of our bones is limited by our sex hormones. A good burst of sex hormones at the right time tells the ends of our bones to stop growing. In acromegalic gigantism, as the tumour grows it destroys cells in the pituitary gland that stimulate the release of sex hormones. The bones, therefore, never get the signal to stop growing.

But surely there must be a limit to a person's height? John Wass, a specialist in acromegalic gigantism at the University of Oxford, reckons it would be impressive to survive for long if you grew taller than 9 ft.

First, high blood pressure in the legs, caused by the sheer volume of blood in the arteries, can burst blood vessels and cause varicose ulcers. An infection of just such an ulcer eventually killed Wadlow.

With modern antibiotics, ulcers are less of an issue now, and most people with acromegalic gigantism eventually die because of complications from heart problems. 'Keeping the blood going round such an enormous circulation becomes a huge strain for the heart,' says Wass.

How long can hair grow?

Hair follicles on the scalp rarely push out more than 0.5 mm of new hair fibre a day and a follicle is active for at most six years before falling dormant. After a few months, it re-activates itself and produces a new hair.

Vietnamese man Tran Van Hay has 6.2 m of the stuff at the time of going to press, although its length may be due to infrequent washing – he has not washed his for six years. 'Hair produces oils and can easily become matted. If you don't wash it, hairs that would have fallen out may stick to those still attached to the scalp,' says Mike Philpott, head of the hair biology research group at Queen Mary, University of London.

Some animals, like angora rabbits, have exceptionally long hair because a mutation in a gene called FGF5 causes hair follicles to be locked into the growth phase for longer. 'Maybe this guy also has a defective gene,' says Philpott. The existing world record, held by Hoo Sateow of Thailand, currently stands at 5.15 m.

Why do fair-skinned Brits burn while Swedes tan?

People from further north tend to have paler skins, the better to absorb the weak sunlight and trigger vitamin D production.

After that any subtle differences in skin type are a matter of genetic inheritance.

'Your Celtic phenotypes – Brits with pale skin, freckles, red hair – will burn and never tan,' says Mark Birch-Machin, reader in molecular dermatology at the Newcastle University and a researcher for Cancer Research UK.

Brits of a less Celtic extraction may burn and then tan when young, but will pay for it heavily with wrinkles when older. 'Each time you go out in the sun and get burned, you damage your DNA. Even before you get sunburned skin, you have damaged your DNA, so it is worse than it looks. You cannot say: "I am safe until I become a lobster." That is not true.'

But Birch-Machin is dubious about races such as the Swedes having any real advantage over us in the tanning stakes. After all, our blood is extremely muddled up in Europe, and the British public is generally exposed to only a small sample of (famous) Swedes – some of whom may sport artificial tans of course.

'If you go out in the sun you may get skin cancer,' he says. 'But what is sure is that your face is going to look like an old sofa. You will have a 50-year-old face on a 30-year-old body, and particularly if you smoke.'

James Scott, director of the genetics and genomics research institute at Imperial College London, thinks that from a genetic perspective, the British should be more likely to toast to a gentle brown than their cousins from more northerly latitudes.

The genetic differences among northern Europeans are minuscule, he says, and any golden glow from the Baltic could be, he says, an 'observer artefact'.

But he is not certain of that. 'Either the genetics is subtly different in Swedes, such that they have blond hair and fair skin but the propensity to develop more melanin when they see the sun,' Professor Scott says. '[Or] maybe there is a form of conditioning in which the genes get set by environmental triggers in a particular sort of way.'

How do you test someone's intelligence?

There are endless methods, each one claiming to have an edge over the others.

Mensa, the UK's high IQ society, prefers to use the Cattel test developed by psychologists in the early 20th century. It avoids using questions that require previous knowledge and tries to measure how quickly and clearly someone thinks. But is it better than the Haselbauer–Dickheiser test for Exceptional Intelligence where each question in the test is a puzzle and the more questions you answer, the more intelligent you are?

'We would say so,' says a spokesperson for Mensa. 'Because it's measuring your speed of thought, which is very important in IQ testing.'

Munder Adahami, a researcher at the Centre for the Advancement of Thinking, King's College London, says that both tests have flaws. 'The problem with IQ tests is that they can be taught,' he says. 'You improve by 10 points by having some practice on them.' In addition, he says, someone's cultural background has an impact on how they interpret, and perform on the test.

Adahami uses the Jean Piaget technique. 'Intelligence is

neither a fixed or inherited quality nor is it something you acquire by experience alone. There's some dynamic interaction between the two.'

It is that interaction the Jean Piaget test tries to tease out. The test does not require any previous knowledge and can eliminate the problems associated with cultural references.

But perhaps the biggest problem in measuring intelligence is actually defining what intelligence is. Many argue, for example, that there is a central processor somewhere in the brain governing our ability to interpret the world around us. Others say this function is spread across different parts of the brain. Working out who is right or wrong is enough to test anyone's head.

Does dyslexia exist?

Not according to some education experts. Instead, they argue, dyslexia is an emotional construct used, in many cases, to save children who are poor readers from embarrassment.

Unsurprisingly, scientists studying the biological basis of dyslexia beg to differ. 'To say it's a myth is pretty far-fetched,' says Tony Monaco, head of neurogenetics at the University of Oxford and an expert on the condition.

According to the professor, children who are simply poor readers may mistakenly be diagnosed with dyslexia if their reading ability is not assessed alongside their general intelligence. The sign of real dyslexia is a reading ability far below that for a child's age and intelligence.

Research is gradually teasing out the developmental glitches that give rise to dyslexia. 'From studies of twins in the UK and Colorado, we know that around 50–60% of the variance in reading ability is due to genetic influences,' says Monaco. The condition is highly hereditary with around half of children born to people with dyslexia also developing the condition.

In a study of 300 families, his group identified a gene on chromosome 6 they suspect is strongly linked to dyslexia. The gene is thought to help neurons in the developing brains of babies move to their correct positions. 'When you knock the gene out in rats, you get no movement of the neurons,' says Monaco.

The finding was bolstered by researchers at Cardiff University who independently identified the same gene as a potential factor in dyslexia. 'In the developing brain, neurons have to move to the right level and it appears that a variant of this gene impairs that movement,' says Monaco.

Brain scans carried out by another Oxford University researcher, John Stein of the Dyslexia Research Trust, have shown that people with dyslexia have underactive brains in several key areas associated with reading and vocal word formation. 'The evidence so far points strongly to dyslexics inheriting a genetic trait that means they have impaired neuronal migration,' he says.

Experts believe that other genes will be discovered that also contribute to a person's susceptibility to being dyslexic. Already, a Finnish group has found a gene on chromosome 15 that impairs neuronal movement in developing humans. And Monaco's group believes that another contributing gene lies on chromosome 18.

Other research supports the notion that it is a real neurological condition: post-mortem examination of brains of people with dyslexia revealed many neurons were in the wrong place.

Why do suicide rates peak in the spring?

Psychiatrists have been scratching their chins over this one for years. Counterintuitively, the arrival of spring, and the long sunny days it ushers in, mark a staggering rise in suicide rates.

Mental health experts at the Priory group say that May is the peak month for suicides in Britain. 'The increase can be dramatic, with up to 50% more successful suicides in some cases,' says Chris Thompson, director of health care at the Priory group. In Britain, about 6,300 people take their own lives each year, 90% of whom are likely to have mental health problems.

The seasonal effect is seen all over the world, with the northern hemisphere witnessing a big rise in suicides in May and June and the southern hemisphere seeing a similar rise in November. While no one has a complete explanation as to why, the leading theory is that the increase is down to the effects of sunlight on our hormones.

According to Thompson, the seasonal changes that bring most of us out of winter apathy may work against those who are coming out of severe depression. 'It is a harsh irony that the partial remission which most depression sufferers experience in the spring often provides the boost of energy required for executing a suicide plan,' he says. 'Spring is a time for new beginnings and new life, yet the juxtaposition between a literally blooming world and the barren inner life of the clinically depressed is often too much for them to bear.'

Paradoxically, says Thompson, sunlight-driven changes in levels of the feel-good chemical serotonin may make people more aggressive and, if they are depressed, they could direct that aggression at themselves. The theory gains some support from research by Canadian scientists linking seasonal changes in bright sunlight with more violent suicides.

Other researchers believe that the influence of sunlight on another hormone, melatonin, is to blame. Sunlight inhibits production of melatonin, which is known to influence our behaviour.

Why do people sleepwalk?

'The bottom line is the brain doesn't move from one sleep state to another properly,' says Carl Hunt, director of the National Centre on Sleep Disorders in Bethesda, Maryland.

'In most of us, when we go into REM sleep, the period when our brain is most active, our muscles are deactivated. It's why we don't act out our dreams. But with these people, their muscles aren't shut down.'

Sleepwalkers can not only walk around, but cook, eat, drive and commit acts of extreme violence. Hunt says sleepwalkers aren't conscious of what they are doing, but because their dreams are influenced by their surroundings, they can get around houses, or even mow lawns naked, as one Ian Armstrong's wife discovered to her horror.

Sleepwalking is common among children, but usually drops off as they reach adolescence. About 4% of adults continue to sleepwalk into older age and about 0.5% become violent in their sleep. The condition is hereditary and more common among men. Although no single neurochemical problem underlies all cases, Hunt says a substantial number go on to develop Parkinson's disease, suggesting that the disorder may be part of a progressive neural condition.

What is persistent sexual arousal syndrome?

There is nothing remotely amusing about this unusual condition. Here's an extract from the internet diary of one anonymous sufferer: 'It continues to rule my life and I schedule my work and personal life around my physical pain and discomfort . . . I have begun to contemplate suicide again because I cannot imagine living like this for the rest of my life.'

This woman is one of several dozen who have come forward in the past couple of years complaining of near-constant sexual arousal. The sensation is apparently unrelenting, not associated with feelings of desire, and can lead to spontaneous, repetitive orgasms. 'I can cause an orgasm by the simple act of gently moving my leg up and down,' another sufferer says.

Sandra Leiblum, a psychiatrist at the Robert Wood Johnson medical centre in New Jersey and one of the first scientists to study the condition, says no one has any idea what causes it. 'It's hard to find a single determinant that unites the women in terms of background,' she says. 'People have looked at everything from hormonal or neurological contributions or excessive vascular flow or congestion in the genitals.'

Finding a treatment has proved equally difficult, though Leiblum says some success has been achieved using counselling with one sufferer, and applying a local anaesthetic called lidocaine with another.

'We would like to do some brainwave studies to see if there's a kind of arousal in the brain centrally that doesn't shut off in the way it would in women without the problem,' she says.

Why is it so esay to raed wrods eevn wehn the lteetrs are mdduled up?

The ease might be illusory. Or even ilosruly. Never mind what the web loggers tell you, scrambled words can be hard to read.

One email doing the rounds claims that 'Aoccdrnig to rsceearh at an elingsh uinervtisy, it deosn't mttaer in waht oredr the ltteers in a wrod are, the olny iprmoatnt tihng is taht the frist and lsat ltteer is in the rghit pclae. The rset can be a toatl mses and you can sitll raed it wouthit graet porbelm. Tihs is bcuseae we do not raed ervey lteter by itslef but the wrod as a wlohe.'

Not so, says Martin Turner of the Dyslexia Institute. 'There is a spectrum of truth here, and that is towards the lower end, because actually sequence is about the only thing that is important.'

Experiments with so-called format distortion can change the appearance of a word drastically – alternating letters in capitals, lower case, superscript and subscript, for instance, or in a huge Gothic typeface to disguise the lettering – but in experiments young children can still read such disguised words, says Turner. What throws them is a change in the sequence of letters, hardly surprising because letters represent a flow of speech sound. The first letter is an important clue to a scrambled word, the last much less so.

In fact, the exact way in which the letters are scrambled can be extremely significant. For example, with plurals, leaving the 's' at the end, but not the letter that should have preceded it, can make the word hard to decipher.

'All you need to do is try and read that email,' says Turner. 'Immediately, you discover it is quite difficult to read. And secondly, you get very fed up with it after two or three sentences. What you have done is put yourself in the position of a dyslexic or poor reader, who loses interest jolly quickly.

'Motivation slumps and it is quite an aversive experience. I got that email, from a fellow psychologist, needless to say, and immediately wrote back commenting that it was hard work, and aversive. After a while, I thought: do I really want to do this? Why don't I look out the window and see what is going on?'

Can you drill a hole through your head and survive?

Yes, though it's not painless. And it depends which bit of your brain you drill through. It was reported, for example, that Ron Hunt of Truckee, California, fell off a ladder and on to a drill, whose 18-inch bit was driven into his skull by his right eye socket and out again by his right ear. Nevertheless, he was laughing and joking with hospital staff shortly afterwards.

'This type of incident is by no means infrequent,' says Steven Rose, director of the brain and behaviour research group at the Open University. He cites the famous case of Phineas Gage, a US railway worker who in 1848 was involved in an accident that resulted in a spike being shot through his head. Gage lived for years but there was a marked change in his behaviour and personality. The case improved our knowledge of where brain functions are localised.

We learned more in the First World War, says neuro-anatomist David Edgart at Liverpool University. 'Bullet wounds produced clean lesions, due to their speed. Doctors could correlate which part of the brain was damaged with the areas of the patient's system that no longer worked properly.'

'To put it in its simplest form, different bits do different things,' says Edgar. 'Physical damage to one part may be fatal, but in another it may have very little effect.'

Rose adds: 'If the lower regions of the brain or spinal cord are damaged – regions that control heart rate, breathing etc. – the consequences are likely to be fatal. The function of the great frontal lobes is more interesting . . . if these parts are damaged the victim is not fatally impaired.'

Why are we happier when the sun is out?

A host of reasons. But, to be fair, none has been proven beyond doubt. The hormone melatonin is believed to play a major role.

When it gets dark, a region of the brain called the pineal gland starts producing melatonin. This is thought to make our bodies cool down and feel drowsy, helping us fall asleep. But flick on the lights and melatonin production is cut off. The 'hormone of darkness', as it is known, does not just make us sleepy. It has also been linked to depression. People who live in regions with very little sunlight tend to have higher levels of melatonin and are more likely to suffer from depression.

In 1997, in an attempt to bring joy to the miserably light-starved people of Helsinki, psychiatrist Timo Partonen of the National Public Health Institute gave people special lamps producing light that closely matched sunlight. After leaving the lamps on their desks throughout the winter people felt happier, less hostile and more alert.

Light also triggers changes in the brain that make us feel more cheery. Evidence is emerging that light pushes up levels of serotonin and noradrenaline, two key feel-good chemicals.

People suffering from seasonal affective disorder (SAD), a mild depression during winter, often crave foods like chocolate and strawberries, says Anne Farmer of the Institute of Psychiatry in London, which are high in tryptophan, a natural precursor to serotonin.

According to Partonen, physiological changes are just part of the story. 'Light has been associated with good, and dark with bad. So, there is clearly a psychological influence,' he says.

What can psychometric tests tell you about someone?

According to its fans at least, psychometrics can go a long way to telling you about (among other things) someone's numeracy, language skills and aptitude for leadership. The tests have become increasingly popular as a method for companies to assess candidates for top jobs.

'You can find out to what degree somebody is likely to be an authoritative rather than a participative manager. You can tell to what degree they believe in learning and development for their team. You can tell a lot about their ability to think strategically,' says Heather Salway, director of the human resources consultancy at Eden Brown Recruitment.

She says psychometrics – a set of tests developed by psychologists and linguists over the past few decades – can help employers learn about their prospective employees very quickly. In under an hour, perhaps, you may learn what would have taken six months to emerge under normal circumstances.

Anyone thinking of cheating would, proponents say, be wasting their time. 'Built into the tests are safeguards so if somebody is trying to fool the tests, that is thrown out when you do the analysis,' says Salway.

Not everyone is enamoured with the tests, however. Their use in hiring staff has been criticised by some trade unions, for example, who argue that they are a blunt instrument likely to encourage laziness among employers.

Can loud music cause physical damage?

Undoubtedly. Turn up the bass, and the high-pressure sound waves can literally knock the wind out of you, causing your lungs to collapse. The condition, known as pneumothorax, was experienced by one man while driving. Doctors blamed the injury on a 1 kW bass box he had installed to boost his in-car stereo. Another man described a sudden sharp pain in his lung while standing next to a loudspeaker at a club.

Doctors believe that tiny pockets of air become trapped in the outer tissues of the lungs, and when hit by intense pulses of sound these air pockets resonate so much they can rupture the tissue, allowing air to leak from the lung. John Harvey, a lung specialist at Southmead hospital, Bristol, teamed up with colleagues in Belgium to highlight the danger. 'In the worst case scenario, the leak is big and continues, so the lung can collapse,' he says. Only certain frequencies, the bass tones between 30 and 150 Hz, are thought to be problematic.

Collapsed lungs are three times as common in men than women, partly because men tend to be taller, so the strain on their lungs due to gravity is greater. Smokers are also more likely to suffer the condition.

No one knows how many collapsed lungs are down to loud music. 'It's never been described before so we don't know how common it is,' says Harvey. 'But whenever we mention it at meetings, people say, "I had a case like that". Now that it's a described association, hopefully doctors will ask patients and we'll probably find it's not uncommon.'

Is urban cycling bad for your heart?

Shedloads of lycra-clad peddlers nearly careered off the road after being told their daily exertion on two wheels could be doing more harm than good. Tiny specks of air pollution belched from diesel-fuelled taxis and buses can damage blood vessels, and, according to reports, could outweigh the obvious health benefits of cycling. The warning came after research in the laboratory of David Newby, a British Heart Foundation senior lecturer in cardiology at Edinburgh University. In his tests, 15 healthy men cycled on exercise bikes in a chamber while being exposed to levels of diesel pollution similar to those found on a congested city street. After an hour of cycling, the scientists found their blood vessels became less flexible and produced less of a protein that breaks down blood clots in the heart – damage associated with the early stages of heart disease.

All very worrying, particularly that, as the faster a cyclist pedals the more air they breathe in, those who believe they are improving their fitness the most are storing up the biggest trouble for themselves.

'Cycling is a good thing to do,' Dr Newby insists. 'We don't want to give the message that it isn't.' His team has made no direct comparison of the risks versus the benefits, and what the original story failed to mention, he says, is that the situation for car drivers is even worse. Drivers face a bigger risk as car ventilation systems hoover up pollution at ground level, which cannot then disperse, leading to concentrations inside some three times higher than on the street.

For cyclists there is both good news and bad news. The latter is that no type of expensive mask and filter can block the tiny carbon particles that do the damage. 'They're just so small they behave almost like a gas,' Dr Newby says.

The positive spin is that the levels of pollution drop off considerably just a few metres from the congested source – so cyclists who stick to less busy roads should have nothing to worry about.

Are sunbeds worse for you than sunbathing?

Given the different types of sunbed available and variations in the intensity of natural sunlight, it's almost impossible to make a direct comparison. But what is clear is that both tanning methods expose the skin to potentially harmful levels of ultraviolet (UV) radiation. 'The simple answer is that sunbeds are about the same as being in the sun,' says Mark Birch-Machin, a skin cancer expert at Newcastle University.

Birch-Machin says results from his laboratory show that exposing cultured skin cells to tanning lamps can induce the same sort of DNA damage that is seen in skin cancer patients. Such DNA damage can lead to potentially fatal skin.

At the heart of the matter is the proportion of UVA to UVB radiation that the sunbeds use. Natural sunlight can contain as much as 8% UVB, which is more intense than the UVA that makes up the rest. Older sunbeds used about 98% UVA, until a succession of orange faces convinced manufacturers they needed to boost the UVB level to mimic sunlight and give a more natural-looking tan. Modern tanning lamps now use around 5% UVB, so they can both tan and burn the skin more quickly, but on the plus side it means that people generally stay underneath them for a much shorter time. Put simply, the new tanning lamps are about as strong as the Mediterranean sun, while older sunbeds are more similar to a day out in Blackpool.

'It's a very unpopular message but a tan is a sign of skin damage because it means your skin has been exposed to UV,' says Sara Hiom of the charity Cancer Research UK. Returning to the original question, Hiom says that it is difficult to judge exactly whether sunbathing or sunbeds carry greater risks. 'To do comparative studies between sole sunbed use and sole solar UV is next to impossible,' she says. 'What people are there that use sunbeds and don't go out in the sun as well?'

Can video games trigger violent behaviour?

Scientists are divided. After decades of arguments over possible copycat aggression sparked by violent films and television programmes, the debate has moved on to the PlayStation generation. The games are certainly violent. Doom 3, for example, features decapitations, exploding heads and stomachs, and an array of terrible weapons including axes, chainsaws and rocket launchers. The British Board of Film Classification said, however, that there was no evidence directly linking the playing of games with violent behaviour.

But some disagree, most notably Craig Anderson, a psychologist at Iowa State University who has published a succession of studies making that link. 'Violent video games are significantly associated with increased aggressive behaviour,' he says. 'High levels of violent video game exposure have been linked to delinquency, fighting at school and violent criminal behaviour.'

Anderson reached his conclusions partly after carrying out his own experiments, during one of which students played violent games and then competed for the right to blast an opponent with a loud noise. He found that aggressive behaviour almost invariably followed aggressive games. He also assessed the evidence of a number of published studies that suggested a link.

The problem is that other psychologists have examined the same studies and reached an entirely different conclusion. 'The research evidence is not supportive of a major public concern that violent video games lead to real-life violence,' one group decided.

How can this be? Partly because it is difficult to compare the studies – the basic assumptions they make are often so different. One, for example, classed Pacman as containing violent events, and counted 0.59 deaths per minute in a Smurfs game.

How do US troops keep going on so little sleep?

Drugs – at least in some cases. American pilots have long been taking them to keep flying longer. The ground troops are thought to be taking a sleep-busting drug called modafinil (or provigil). Away from the battlefield, it is used mainly to combat narcolepsy, a condition that makes people feel irresistibly drowsy. The ears of military commanders pricked up when they heard it could keep a healthy person awake for more than three days. The Americans, British and French have all tested it.

A US study found sleep-deprived helicopter pilots given modafinil were more alert, energetic and confident than those who were not. A French researcher, Michel Jouvet, of Claude Bernard University in Lyon, concluded that the drug could keep an army fighting for three days and nights.

According to its US manufacturers, Cephalon, the only side-effect is a headache, which wears off when the user gets used to the drug. Exactly how the drug works is unknown but it is believed to stimulate a part of the brain called the pre-frontal cortex, used in planning and problem solving.

Officially, neither the Americans nor the British armed forces sanction the use of modafinil. But a source at Cephalon said he would be 'astonished' if troops were not using it.

The need to sleep has held armies back throughout history. Other 'fatigue management tools', as the American military puts it, are used with varying degrees of success.

Caffeine is the preferred substance for many but since the Second World War, US military pilots have been authorised to pop dextroamphetamine, or 'go pills', which have been under the spotlight since the 'friendly fire' killings of four Canadian soldiers in Afghanistan.

The search for the 'no-doze soldier' continues: researchers at the US Defense Advanced Research Projects Agency are looking at how birds and dolphins can survive on little or no sleep. So, soldiers of the future may wonder why the charge through Iraq was so exhausting for their predecessors.

What age should we work until?

If reports are anything to go by, workers in the year 2050 will be waiting until the age of 68 to retire. But how old is too old?

Time was when neuroscientists had a simplistic view of how our brains decline with age, eventually making us incapable of anything but the most unchallenging of jobs. 'The idea was that we were born with zero cognition, died with zero, and peaked somewhere between 25 and 30 years old, after which it was all downhill,' said Itiel Dror, an expert in cognitive neuroscience at Southampton University.

Researchers now know that while brain cells die off with age, the remainder make new connections to compensate. According to the American Federation for Ageing, brainpower is at its best in the thirties and forties, with a drop only becoming noticeable in the seventies. One of its studies found that by the age of 81, two-thirds of people showed only a minor decline in cognitive ability.

Dr Dror says the variation in cognitive ability is so large that it is impossible to give one age when everyone is too old to work. 'It depends so much on the job, but why say everyone has to retire in their late 60s? Some people will be in no shape to work long before that, while others will be fine into their 80s,' he said.

Among artists, Picasso continued working into his 90s and Goethe knocked out *Faust* as an octogenarian.

The annoying part for those who want to retire young and do nothing, says Dr Dror, is that without work or other types of mental stimulation, our cognitive decline is much steeper. 'It's a case of use it or lose it,' he said.

Sickness & Health

How long before I will be able to take personalised medicine?

'The dream of the enthusiasts would be that you would be able to tailor medications to those people who were either likely to get the biggest efficacy response or to have the minimum side-effect response,' says Patrick Vallance, head of the school of medicine at University College London. 'If you had pain, you'd know which drug was the best for you.'

The most immediate hope is that pharmacogenetics will give doctors more information to prescribe drugs that will cause minimal side-effects in their patients. The fact that people respond differently to the same medicine can be explained by minor variations in their genes.

A gene contains a coding part and a non-coding part. The former encodes for the various proteins that will make up, say, a receptor to a certain drug. A mutation here could cause disease, or may even be lethal.

But if there are minor variations in the non-coding part of

the gene, it might have more subtle effects. Vallance says: 'It may affect the way the drug is handled in the body.'

One promising area for pharmacogenetics is connected to drugs that have failed clinical tests. 'Some drugs fall by the wayside because they cause a nasty effect in a small number of people,' says Vallance. If scientists could work out which genetic profiles predisposed people to those side-effects, the drugs could be targeted to people for whom they are safe.

But in some cases, pharmacogenetics could be overkill. Vallance says it might end up being better to treat somebody for hypertension by trying out one drug, measuring any changes and then trying out a different drug if the first one hasn't reduced blood pressure. 'You may not get much further by starting off with a very fancy genetic profile and a specific drug,' he says.

Will antibiotics be useless by 2015?

It's a distinct possibility – over-use of these drugs has certainly led to an increased number of resistant bugs. The worry is that, one day soon, a superbug will emerge that is resistant to all forms of antibiotics.

As bacteria multiply, random mutations can make some of them invulnerable to certain antibiotics. Once one bug is protected, the resistance can quickly spread through a population and eventually render certain drugs useless. Biologists have therefore spent the past few decades regularly developing new antibiotics in a race to beat the evolving bacteria.

According to scientists at Ulster University the situation has been exacerbated by the over-prescribing of antibiotics by doctors and vets in the past 50 years. As more bacteria get exposed to the drugs, more become resistant. The scientists say this over-use will lead to all antibiotics becoming useless in just over a decade.

'There are very significant problems in hospitals with infections with a number of organisms which have become very difficult to treat,' says Brian Spratt, a microbiologist at Imperial College London.

The most well-known superbug is methicillin-resistant *Staphylococcus Aureus* (MRSA), which can cause anything from skin infections to septicaemia or pneumonia. The bug has been common in hospitals for around a decade but can be controlled by only one antibiotic. That antibiotic is used sparingly, but it too will, one day, become useless, which will mean that even small cuts infected with MRSA could become untreatable.

Making new antibiotics is the answer. It's not easy, though, and can take anything up to 15 years of research, development and clinical testing. Some scientists, therefore, are looking at other ways of combating bacteria.

The most promising alternative are bacteriophages – viruses that can infect and kill bacteria. They occur naturally, and the trick is to try to isolate them from the environment for use in healthcare.

'Bacteriophages have been around for an awfully long time,' says Nick Hounsby, a biochemist and chief executive of Novolytics, one of only two companies in the UK actively involved in developing phages. 'Before antibiotics became apparent, they were going to be a very useful treatment. Because antibiotics came along, and were so easy to give to people, they got forgotten.'

The good news is that they can be developed quickly – after the bacteria in a wound has been isolated, a bacteriophage cream could be produced within weeks.

Spratt is not convinced: 'I don't think they're considered to be credible alternatives by the great majority of people. My suspicion is that if you started using that type of approach in the real world, it might work a little bit but you'd very soon have problems with the bacteria becoming insensitive to the virus.'

Can the wind cause outbreaks of meningitis?

Indirectly, for people in a narrow band of West Africa. Scientists from the French Institute for Research and Development (IRD) and colleagues at the University of Paris have solved a conundrum that has puzzled public health experts. At the same time each year, a meningitis epidemic strikes up to 200,000 people, particularly children, in Mali and the surrounding region. The disease spreads only in latitudes between 10 and 15 degrees north, the so-called meningitis belt.

Scientists suspected the weather was involved. Now the IRD experts have named the culprit as warm and dusty winter winds called the Harmattan. The northern winds damage mucous membranes in the throat and lungs, helping the meningococcal meningitis bacteria get into people's bloodstream.

Records from 1994 to 2002 show that the Harmattan blows

hardest in the second week of February, which tallies with the onset of the annual outbreak. When the wind dies at the onset of the rainy season in mid-April, infections tail off.

How can you protect yourself from bird flu?

In Asia, the most worrying strain of the virus, H5N1, has spread through poultry, triggering countries to order mass cullings. Since 2003, the virus has killed more than 130 people, almost all of whom are believed to have picked it up from infected birds.

There are measures we can take to limit our chances of picking up the virus if it does reach Britain, says John Oxford, an influenza expert at Queen Mary, University of London.

'Regular handwashing with soap, which seems very old-fashioned, could actually make quite a big difference,' he says.

The loss of handkerchiefs to tissues, he says, means that people blowing their noses are more likely to end up with live virus on their hands and this can then be transferred to surfaces. 'The kinds of surfaces you'd want to avoid touching or clean well would be door handles, keyboards and phones, anything that is shared,' he continues.

Another obvious but effective strategy is what scientists call social distances – staying away from anyone else who's got the virus. Trains and buses are obvious places where an airborne virus could spread, but football matches and workplaces are also high on the list. 'You'd also want to stay away from GP's surgeries and hospital A&E departments because they are where people with flu will be congregating,' says Oxford.

Many scientists believe too little is known about the immune system to recommend supplements that could usefully boost it.

The virus that caused the 1918 flu pandemic was most lethal to 25- to 35-year-olds. 'These people tend to have immune systems at peak condition and there's more than a suggestion that in 1918, many died because their immune system over-reacted to the virus,' says Oxford.

What is the best thing to wash your hands with?

Depends on what you are worried about: bacteria or viruses. The answer to this question might be simpler than you had thought. Emily Sickbert-Bennett and colleagues at the University of North Carolina noticed that healthcare workers in their institution spent about 10 seconds on average cleaning their hands, rather than the 30 seconds or so assumed by most previous research.

The researchers looked at how well 10 seconds of scrubbing with various things, from tap water to the alcohol hand-rubs introduced by the government to British hospitals, removed the bacterium *Serratia Marcescens*, or a virus called MS2, from contaminated hands. The results, published in the *American Journal of Infection Control*, showed that the best way to get rid of bacteria in 10 seconds was with an antibacterial soap, whereas the virus was sent packing with just regular tap water. This physically knocks the particles off the hands and down the basin.

Lots of people are interested in handwashing these days,

because it has been shown over and over to cut rates of hospital-acquired infections, such as the 'superbug' MRSA. This is universally agreed to be a good thing. As for alcohol-based rubs? Well, they did fine at first, but with repeated cycles of contamination they fell behind good old soap and water.

The scientists suggest that healthcare workers should wash their hands the old-fashioned way, as well as using the new-fangled alcoholic gel rubs. But whatever they use, spending more than 10 seconds at it might be quite helpful too.

Can popping pills protect you from radiation?

It certainly helps. By stopping your body from absorbing radioactive iodine (also known as radioiodine) churned out as a by-product in nuclear power plants and atomic bombs, potassium iodate pills have been invaluable for people who have been exposed.

Radioiodine is particularly dangerous because it is so easily absorbed. 'The almost unique property of the thyroid gland is that it's the only organ in the body that is so hell-bent on absorbing one particular element,' says Stanley Bachelor, a radiation protection adviser at King's College London.

Even very small amounts of inhaled or ingested radioiodine can cause damage because it can concentrate in the thyroid gland, which uses iodine to make hormones. Once in the body, radioiodine can lead to loss of thyroid function and, eventually, cancer.

Taking potassium iodate pills floods the body with non-radioactive iodine and allows the thyroid to fill up with the element. If exposed to radioiodine, the body will simply absorb less.

The effectiveness of the chemical depends on how much you take, according to Bachelor: 100 mg (a typical dose) can reduce uptake of radioiodine by more than 50%.

Potassium iodate will not protect anyone from other radioactive ingredients that may be found in a dirty bomb such as cobalt, caesium or uranium. These are not concentrated so readily by the body's organs.

Is suspended animation safe?

If you need it, the safety of the procedure will probably be the last thing on your mind. Researchers are exploring the idea of 'suspended animation' to help save the lives of severely injured people, who would be likely to die without it.

Funded by the US military, scientists at the Safar centre for resuscitation research at the University of Pittsburgh succeeded in reviving dogs who had been put into the zombie-like state for two hours. The dogs had their blood drained from a vein, which eventually led to their hearts stopping. A cold saline solution was then pumped into their blood vessels, keeping their body close to 7°C instead of the usual 37°C, effectively lowering metabolism. The dogs were resuscitated by pumping the blood back into their veins and giving them electric shocks to the heart.

Rather than being interested in the technique to prevent future astronauts dying from old age or boredom on long-haul missions – Ripley enjoyed a state of hypersleep in the

1979 film *Alien* – the research might one day help save those with serious wounds. By putting people into suspended animation, the researchers hope to give medics more time to operate on patients before their condition deteriorates.

The procedure is extremely risky, not least because of the danger of insufficient oxygen getting to the brain. In the research leading up to the latest work, several animals suffered neurological damage after being resuscitated. In one experiment, the Safar team used the technique to resuscitated dogs one hour after being given heart attacks. The experiment was designed to mimic more closely real trauma cases and the dogs did not suffer brain damage.

Mark Roth, at the Fred Hutchinson cancer research centre in Seattle, reported in the journal *Science* that hydrogen sulphide could also be used to force creatures into a state of suspended animation. His tests on mice found that when exposed to the gas, their metabolsim dropped and they fell into a hibernation-like state from which they later recovered, apparently unharmed.

How much can a coma patient understand?

Pretty much everything according to Salvatore Crisafulli, a 38-year-old Italian man who awoke after a two-year coma following a road accident in 2003. He claimed to have 'understood everything and cried in desperation' while trapped inside his body. But Ronald Cranford, a neurologist at the Centre for Bioethics at the University of Minnesota in the US, who testified in the case of Terri Schiavo – the brain-damaged

Florida woman who died when her feeding tube was removed – is not convinced. 'This story seems to be a first-person account of a "miracle recovery",' he says.

David Bates, a professor of clinical neurology at the Royal Infirmary in Newcastle, thinks that it is more likely that Mr Crisafulli was suffering from 'locked-in syndrome'. 'This is a condition where all the information going into the brain continues as normal, but all the information coming out stops. The patient can hear, feel, see and understand, but they can do nothing, except open and close their eyes,' he explains. To differentiate between patients in locked-in syndrome and those in a coma, doctors ask the patient to open or close their eyes. Patients with locked-in syndrome will be able to open and close their eyes voluntarily and respond to the doctor's command.

It is unlikely that Mr Crisafulli's case alone will alter the way doctors diagnose and treat unconscious patients. But as the Terri Schiavo court case revealed, it is not just experts such as Mr Cranford who need convincing: families also need to be sure that there is no hope of recovery. In the US and Italy religious groups campaign for obligatory care for unconscious patients – even those who had said they did not want extra-ordinary measures to keep them alive.

Is laser eye surgery dangerous?

Not necessarily. Laser assisted in-situ keratomileusis (Lasik) has proved popular among those with short sight and is performed under a local anaesthetic. 'The cornea has a small flap cut into it with a motorised razor blade and that's a predetermined thickness,' says Larry Benjamin, consultant ophthalmologist at Stoke Mandeville hospital. 'That's lifted up and the bed of the cornea is reshaped with the laser and the flap is put back again.'

The cornea does most of the light-bending in the eye so reshaping it will alter the eye's ability to focus light.

Says Adrian Ward, a spokesman for the National Institute for Clinical Excellence (NICE): 'In terms of efficacy for mild to moderate short-sightedness, it was deemed to be efficient but there were concerns about the procedure's long-term safety.' NICE is worried that, in 20 or 30 years, people who had Lasik surgery might have structural problems with their cornea.

'If you're burning out an area, it's not something that can be done over and over again. Not everyone's got the same strength of cornea,' says Ward.

There is also the risk of infection or more extensive damage to the eye. 'The flap never quite sticks fully – you can lift it even years later with fairly minimal surgical intervention,' says Benjamin. A knock to the eye could dislodge the flap completely.

What is awake craniotomy?

Over to Greg Scott, who was diagnosed with a brain tumour in 2000, and kept an internet diary of his treatment. 'When my surgeon first told me on the phone that I was to have an awake craniotomy, I thought wow,' he wrote. 'As soon as I got off the phone I did a quick search on the internet and found two pages of text which described the procedure. I thought wow, no way.'

The technique, which has been around for years as an epilepsy treatment, is rare and only used when a certain type of tumour lies close to specialised areas of the brain that control movement or speech.

Although brain tissue does not register pain, the scalp does, meaning patients are given a general anaesthetic and only brought round when a hole has been cut in the skull. The patients are then asked to talk, count and recognise pictures. As the surgeon reaches the fringes of the tumour, the performance of the patients becomes notably affected.

'One of the difficulties is that if I stop as soon as people start getting worse I'm probably stopping too early,' says Henry Marsh, the only brain surgeon in the country to perform the technique. 'You can introduce a certain amount of mal-function that then recovers. It would be easier if I stopped as soon as they started making mistakes but I know I can push on a bit longer.'

Is it safe to put mercury into vaccines?

It appears so, although the waters have been muddied by Mark Geier, a scientist who runs a private institution, the Genetic Centres of America, from his home in Maryland. Geier claimed that babies given a mercury-containing triple vaccine, to protect against diphtheria, tetanus and whooping cough, were six times more likely to develop autism than those given a mercury-free alternative.

Mercury, in the form of a compound called thiomersal, has been added to a handful of different vaccines since the 1930s to prevent them from spoiling. While thiomersal is mostly found in combined diphtheria and tetanus vaccines, it is not added to MMR, polio or the BCG vaccine for tuberculosis.

Scientists have long known that mercury is toxic to brain tissue. When cases of autism were found to be increasing around the world, some blamed thiomersal.

But most scientists believe thiomersal is safe at the levels used in vaccines. A typical dose of a thiomersal-containing vaccine contains around 25 mg of mercury. 'Mercury is a neurotoxin, so it does raise questions, but water is a neurotoxin too at high enough doses. If you're going to talk poison, you have to talk dose,' says Karin Nelson, a child neurologist at the National Institute of Neurological Disorders and Stroke in Maryland. Nelson does not believe thiomersal causes autism, not least because the symptoms of mercury poisoning are quite different from those of autism. 'No study has ever shown that children exposed to mercury in vaccines, or by any other route . . . have more autism than children without such exposure,' she says.

Is heroin safe for some people to use?

Perhaps, say researchers at Glasgow Caledonian University, who surveyed 126 long-term users who seemed to have none of the health or social problems associated with the drug. Definitely not, responded some anti-drug campaigners, who complained that the research was irresponsible.

In any case, the results are not entirely surprising for people working in the drugs field: research from America in the past few decades has documented several instances of people who can use drugs without spiralling off into so-called chaotic use.

Previous work found that people who managed their drug use often set themselves a strict framework to operate in. For example, some would never use it more than two days on the run or they would avoid using it on a Sunday night, if they were working the next day. Most of the users surveyed in Glasgow were in employment, none had been to jail and they were well educated. They had plenty of social support and most were in relationships or had families around them.

Is cannabis addictive?

There is evidence to suggest that is is psychologically, if not physiologically.

Michael Rowlands, medical director at the Priory Farm Place, says cannabis shows all the classic signs of dependency. 'There's a strong desire to use, which overrides other activities, so friends and hobbies and work are neglected,' he says. 'There's difficulty in controlling the amounts you use. There's a degree of tolerance developed so you need higher doses to have the same effect. And then you persist in using despite the fact it's causing you ill health or debt.'

The main thing that separates cannabis from heroin or nicotine is that the physical withdrawal state is not normally as severe.

Almost all addictive drugs stimulate a part of the brain –

called the mesolymbic dopamine system – that acts as a reward pathway in the central nervous system. Receptors for the active ingredients in cannabis have been found in this system. Once stimulated, these receptors begin a cycle of reward that can lead people to use more of the drug.

Rowlands says an apparent increase in cases of addiction might be nothing more than a product of the changing attitudes towards cannabis use. 'Some of the stigma is going. People are much easier at talking about addiction,' he says. 'There are vast numbers of people taking cannabis. Some of them, 8 to 10%, will get some type of dependency.'

More concerning than any apparent rise in addiction is the potential to cause psychoses in heavy users.

Robin Murray, a psychiatrist at King's College London, is one of Britain's leading researchers in this area and his results are worrying. 'The conclusion was that, if you took cannabis at age 18, you were about 60% more likely to go psychotic. But if you started by the time you were 15, then the risk was much greater, around 450%,' he says.

What are the real risks of cocaine?

Model Kate Moss may have publicly apologised for her use of the drug, but the long-term damage may go well beyond temporarily lost modelling contracts.

According to John Henry, an expert in illicit drugs at St Mary's Hospital in London, cocaine is undoubtedly dangerous. 'It tightens up all the blood vessels in the body, so the heart has to work much harder, but at the same time its ability to pump harder is reduced, so the whole system is put under extreme stress.'

A survey at St Mary's found that between 7% and 10% of people arriving at the hospital's accident and emergency centre with chest pains had traces of cocaine in their urine. For the under-forties, that figure rose to nearly one third or as high as 50% on Saturday night. 'We did a control experiment by testing the urine of others in A&E without chest pains and still 3% of those had traces of cocaine in their urine,' said Professor Henry.

US studies suggest the chest pains caused by cocaine are not benign. Around 5% of people turning up at hospitals with such pains go on to have heart attacks. Others have strokes as the rise in blood pressure bursts blood vessels in their brains. There was also evidence that long-term cocaine use damages brain function, added Professor Henry.

'We see people in their early thirties with severe coronary artery disease, but they're not smokers, not overweight, and don't have high blood pressure. Cocaine is their only risk factor. Some people just cannot control their use.'

Why does LSD make you hallucinate?

The hallucinations happen because the drug mimics a chemical messenger in the brain called serotonin. While serotonin is usually described as a 'feel-good' chemical – it is the neurotransmitter released by the drug ecstasy – it also plays a number of other roles.

The brain has at least 14 different receptors for serotonin, all of which play a different part in regulating functions such as our mood and how we interpret what our senses tell us.

'We think serotonin helps keep a handle on perception and actually stops us from hallucinating,' says Clare Stanford, a psychopharmacologist at University College London.

A dose of LSD, or lysergic acid diethylamide, targets a specific serotonin receptor called 5-HT2A, and in doing so appears to throw our senses into a jumble. As a result, images we would never normally perceive become vivid and fool our brains into thinking they are real.

'The drug can also cause synaesthesia, a condition which happens naturally in a small percentage of the population, where your senses get mixed up and you start smelling colours and tasting sounds,' says Dr Stanford.

How does alcohol protect against disease?

Pretty well, in moderation at least. Results from Denmark have confirmed what scientists have known for some time: moderate consumption of alcohol protects against some of Britain's biggest killers, including heart disease and diabetes.

The study, led by Morten Gronbaek, of the Centre for Alcohol Research in Copenhagen, analysed the health and drinking habits of 57,000 people aged 55–65. He found that people who drank little and often had a significant reduction in mortality. 'The results show that alcohol can be good for your health provided you adopt a careful drinking style,' Gronbaek says.

After decades of debate, this cheering conclusion is now

accepted by most researchers in the field, who have become more interested in unpicking exactly how alcohol brings such benefits. With coronary heart disease, the key seems to be cholesterol, of which there are two types, known as high- and low-density lipoproteins (HDL and LDL).

Crudely, LDL, high levels of which are associated with a greater risk of heart disease, is bad; HDL cholesterol seems to protect the arteries and is good. Alcohol raises levels of this good HDL cholesterol.

'Somewhere in the range of a third to two-thirds of the benefit of alcohol is likely to be related to HDL,' says Kenneth Mukamal, an expert in alcohol and health at Harvard University. Scientists still don't understand how ethanol (the actual alcohol in all booze) has this effect – some speculate that it boosts HDL production, others that it slows the rate at which it breaks down.

Studies have shown that HDL levels steadily rise with up to about four to five units of alcohol consumed a day. 'When you get beyond moderately heavy drinking it goes all over the place,' Mukamal says. 'In some people levels go higher, in some they go down. Certainly we cannot say that higher drinking leads to higher HDL levels the way moderate drinking does.'

On how alcohol might ward off diabetes, the picture is fuzzier. Some clinical trials suggest that it can improve insulin sensitivity, making the body better able to process blood sugar, but it is unclear how that happens.

Does what you drink make any difference to your health? The jury is still out. Some research suggests that red wine is best but many experts remain unconvinced.

'We don't have any evidence that beverage type matters, at least in the main mechanisms of HDL and cholesterol,' Mukamal says.

Does passive smoking kill?

'We have quite a body of evidence building up over the last 20 years,' says Sinead Jones, director of the tobacco control resource centre at the British Medical Association (BMA). This includes a world-wide review of research in the field, pulled together and published by 40 epidemiologists for the World Health Organisation that concluded second-hand smoke increased the risk of lung cancer by 25%.

'One of the most interesting studies is autopsy evidence from wives of men who smoked,' says Jones. 'These women actually died of other, unrelated diseases, but they did lung biopsies and found pre-cancerous changes very similar to those seen in smoking-related disease.'

But Simon Clark, director of the smokers' lobby FOREST, says anti-smoking campaigners have not proved their case. 'All we ever hear are estimates, calculations and statistics,' he said. 'Where is the hard evidence that people are dying of passive smoking?'

And some research published in the *British Medical Journal* even seemed to back FOREST's claims. The universities of California and New York analysed data from more than 35,000 people who had never smoked, but lived with a spouse who did. Their paper seemed to find no link between passive smoking and death from lung cancer or heart disease.

But Jones says the paper had been misinterpreted. 'They [the smoking lobby] are always entitled to say 'we believe that it doesn't cause lung cancer', and I can believe that the Earth is flat. However, believing that doesn't make it so.'

Is it worth detoxing?

At the risk of denting the willpower of anyone struggling with a draconian detox regime, the truth is that it is unlikely to do you much good. For starters, a couple of weeks at Christmas, bingeing on the unholy trinity of turkey, sprouts and alcohol, will have a negligible effect on your long-term health. Your body should be quite capable of making sure you don't suffer too much for your excesses.

'People underestimate the ability of our own bodies to deal with toxins,' says Toni Steer of the MRC Human Nutrition Research Centre in Cambridge. 'The liver and kidneys are extremely sophisticated organs that are capable of detoxifying you without you needing to live on water, orange juice, or whatever the detox regime is.'

Steer says detox diets are often seen by people as a quick fix to get rid of a few pounds. But any weight loss is unlikely to last for long, she says. 'If you're on a detox diet, what you're doing is simply decreasing your calorie intake. If you do that, the first thing the body does is use up glucose in the blood for energy. Then the body has to use up glycogen, a stored form of carbohydrate,' she says. Glycogen is essentially just glucose bound to water molecules and is stored in the liver and muscles. After a day or two of detox, the body begins to break down glycogen, using the glucose from it for energy and releasing the water to make urine. 'People see they are losing a lot of water and think they are flushing out all the toxins, but all they're doing is using up glycogen.' Although the water loss will shed a pound or two, it'll go straight back on as soon as the detox gives way to a more regular diet.

Cutting out alcohol for a long period is also going to do little for your health. Drinking in moderation is healthier than not drinking at all, so simply cutting down to the recommended intake – no more than 14 units a week for women and 21 units for men, where a pint is two units and a glass of wine one, is the best strategy – unless you are simply sick of the taste of the stuff.

Is there a link between breast cancer and antiperspirant?

This scare started in 1999 with a circular email claiming that toxins were 'purged' through perspiration, and that when the armpit sweat glands were blocked, toxins built up in the lymph nodes behind them, causing cancer in the upper outer quadrant of the breast.

While it's true to say that an excess of cancers occurs in this quadrant of breast tissue – the one closest to the lymph nodes – it is also by far the largest area of breast tissue.

Furthermore, sweating is certainly not the primary means by which the body rids itself of toxins. But the email caused so much concern at the time that the American Cancer Society and the National Cancer Institute issued statements to reassure people, and scientists began to study the issue.

If you really want to get to the bottom of whether an environmental factor is causing an illness, it's good to be able to compare one group who were exposed to your toxin, antiperspirant for example, with a group who weren't. In 2002 a large study was published looking at 1,600 women and found no link between antiperspirant use and breast cancer. It was testament to the influence of the original 'hoax' email (as Harvard medics called it) that they also went out of their way to study its specific claim that shaving before using antiperspirant increased the chances of it causing cancer, because toxic chemicals could get in more easily, and again found no link.

Unfortunately, it looks like the largest risk factors for breast cancer at the moment are the ones we can't control such as age, sex, and family history or the ones we might not want to such as smoking and age at childbirth.

What is hair cloning?

The treatment has nothing to do with cloning. It relies instead on the hair-producing capabilities of cells called dermal papillae at the bottom of hair follicles. 'The cells work by recruiting nearby skin cells to make new hair-producing follicles,' says Paul Kemp of Intercytex, a Manchester firm carrying out clinical trials of the technique.

Scientists have known for years that dermal papillae cells from follicles at the back of the head can be injected into a bald scalp to make hair sprout again. But getting the technique to work well has been tough.

The first hurdle, which scientists have now overcome, was working out how to make dermal papillae cells multiply in a dish. Doing this means only a few cells are needed to grow enough to cover the baldest of heads.

The second hurdle scientists faced was how to get the cells into the scalp. Because the cells have to go into such a thin layer of skin, tiny syringes, which inject just a few microlitres, are being used in Manchester. The trial should now be complete, although final approval for the technique is unlikely before 2009.

Kemp says that to give someone as bald as Hollywood star Bruce Willis a new head of hair would take 1,000–2,000 injections, each of which is done by hand. As painful as it sounds, the needles used are so fine, it should be more comfortable than hair transplants, he says. 'Normally, a hair transplant is done one hair follicle at a time. It's painful and time-consuming – it can take 16 hours to do a full head.'

Why do the Japanese live so long?

They eat better, are less stressed and get more exercise. The number of citizens aged at least 100 has doubled in the last five years, making Japan world number one in longevity.

Life expectancy is longest in the world for both sexes: 85.23 years for women, 78.32 for men in 2002.

Greg O'Neill, director of the US National Academy on an Ageing Society, says the only known factors that might prolong health, such as restricting calories, seem to be making a difference. People in Japan eat a third fewer calories than the typical North American. What they eat is also important: more seafood and, hence, healthy fish oils.

Robert Arking, biology professor at Wayne State University, Detroit, says that the Japanese are less sedentary than westerners, and Japan is a less stressed society with more equity in income, status and workplace hierarchy.

O'Neill adds that just after the Second World War, the Japanese had one of the lowest life expectancies in the world. This suggests that increases in lifespan are most probably unrelated to genetics.

Does being obese give you cancer?

There isn't much doubt that there is a link, says Jane Wardle, of Cancer Research UK's health behaviour unit, which released research revealing how little the British know about the costs of obesity. The survey found that 70% were aware of the link between heart disease and obesity. Only 3% made the same connection between fat and cancer.

In fact, says Wardle, it isn't at all clear how the fat–cancer link works but it does work. US doctors monitored a million Americans for 16 years and found that the obese ones were much more likely to develop cancer than the skinny ones. There could be several reasons. The extra weight itself may create a problem. To take one case, obese people might be more liable to what is politely known as gastric reflux (stomach acid rushing back up the gullet) which could create an extra risk of cancer of the oesophagus.

Then again, the risk might lie in the diet of fat people –

which would contain a lower proportion of fresh fruit and vegetables – or it might be that eating less is in itself healthy, which in turn makes eating too much a bad thing. The killer connection might lie in the differing lifestyles of the slender and energetic, and the plump and sluggish, or in the underlying behaviour that caused the obesity in the first place.

The hazards might lie in the fat itself. Doctors used to think of it as a neutral energy store, Wardle says. But fat is a source of hormones – oestrogens and so on – which are biologically active: the greater risk of breast and uterine cancer in post-menopausal overweight women could be linked to these. 'You have a lot more things swilling around in your system,' she says.

There is no doubt that losing weight is good for health in general. But even that exposes a gap in understanding. 'What we don't know for sure is that losing weight will lower the risk of cancer,' she says.

Does banning smoking in public places encourage people to quit?

Well, why do you think the tobacco industry opposes the idea? Research in California – where smokers have been banished from bars, restaurants and offices since 1998 – shows that smoke-free environments reduce both the number of smokers in the population at large and the number of cigarettes they get through. To assess what happens when bans are imposed researchers first study populations that can be tracked. So the Californian study looked at smoke-free workplaces in the US,

Australia, Canada and Germany. It found that smoking levels among employees dropped a whopping 29%. The number of people indulging dropped by nearly 4% and those who kept on smoking puffed their way through an average of three fewer fags per day. To achieve the same reduction using taxes, the researchers estimated that the price of a packet of cigarettes would have to rise by 73%.

'A ban accelerates the rate of cessation because it gives people another reason not to smoke,' says Thomas Houston, a doctor with the American Medical Association. Even internal tobacco company documents estimate that if all work places were smoke-free, overall consumption would drop by 10%.

'It works both ways,' says Joaquin Barnoya of the centre for tobacco control, research and education in San Francisco. 'For non-quitters the number of cigarettes decreases and those that are trying to give up find it easier.'

Tim Lord of the Tobacco Manufacturers' Association argues that the picture is more variable. 'There are some studies where they've seen a reduction in consumption and some where they haven't,' he says.

When was the link between smoking and cancer established?

It seems obvious now that smoking is bad for you. But, back in the first half of the last century, things were different. Medical textbooks were largely empty on the subject, and smoking was often seen as part of growing up.

According to Jean King, director of tobacco control at

Cancer Research UK, the seminal document that linked lung cancer to tobacco and demanded action from the government was published in 1962 by the Royal College of Physicians. And it took time before its effects were widely felt.

Evidence had been gathering for more than a decade beforehand. In 1949, Richard Doll, a researcher working for the Medical Research Council, and Bradford Hill, an epidemiologist at the London School of Hygiene, began looking at lung cancer patients in London hospitals. The patients were asked about family history, diet and previous diseases. In 649 cases of lung cancer, two were non-smokers. Doll immediately gave up his own five cigarettes a day habit.

Doll and Hill extended their research to Cambridge, Bristol and Leeds and, after speaking to some 5,000 people, found the same results.

In 1951, the researchers wrote to 59,600 doctors and asked about their smoking habits. They kept a watch on the doctors' health and published the results in 1954 in a paper now deemed so important that the *British Medical Journal* reprinted the first page 50 years after the original report.

Doll and Hill followed up their work and, by 1956, the link was incontrovertible: more than 200 heavy smokers had died in a four-year period while the incidence among non-smokers was negligible.

After the Royal College's recommendations in 1962 – restriction of advertising, higher taxation, restrictions on sales to children and on smoking in public places, information on tar and nicotine content – cigarette sales fell for the first time in a decade.

Do any complementary medicines work against cancer?

No – as Edzard Ernst, based at the Peninsula Medical School in Exeter and Britain's only professor of complementary medicine, says. The problem is that many websites advocate the use of complementary therapies instead of conventional ones to treat the disease.

Ernst's team analysed 32 of the most popular websites giving advice and information on a range of complementary therapies to treat cancer. Between them, they receive tens of thousands of hits a day. He concluded that a 'significant proportion' of the sites were a risk to patients.

'This was to us quite an eye-opener and pretty scary stuff,' Ernst said. 'Among these 30-odd sites, 118 different "cures" were recommended. None of these 118 can be demonstrated to cure cancer.'

This is not to say that complementary medicines have no place in treating illness. Garlic can lower cholesterol and acupressure is an effective treatment for nausea and vomiting.

Evening primrose oil, for example, is often prescribed by complementary therapists for everything from arthritis to pre-menstural syndrome. Not one scientific study backs up any of its claimed uses.

Can dogs sniff out human cancers?

Although Lassie never saved the day by excitedly leading doctors to a hidden melanoma, stories of cancer-spotting dogs abound. The first involves a border collie–Dobermann cross that in 1989 evidently sniffed out a cancerous mole on a woman's leg. Then, in 1997, George, an explosives-sniffing schnauzer, was trained to sniff out skin cancers. Despite George's reported success, dogs are not yet standard equipment in hospitals. 'This idea still has to be scientifically verified,' says Paul Waggoner, Director of the Canine and Detection Research Institute at Auburn University in Alabama.

Scientists now want to test if dogs can smell the difference between samples of urine from people with cancer and urine from healthy people. It's not a crazy idea, says Waggoner. Many cancers are known to shed specific proteins into the bloodstream that can also make it into urine. If they have a distinct scent and dogs' noses are sensitive enough to pick them up, it might just work.

Medicine has a long history of using smell to diagnose disease and groups at Imperial College London and Cranfield University in Bedfordshire have worked on 'electronic noses' to sniff out infections. Scientists tend to opt for sensors they have built because they are easier to calibrate reliably.

Waggoner says that it might make more sense to work out what cancer proteins can be found in the urine and develop a chemical test for those. An advantage of using dogs, he concedes, is that you don't need to know what the tell-tale protein or other cancer-related chemical they are sniffing is. This could be important if the protein or chemical was at such low levels that machines would not detect it.

What is hot tub lung?

If you're reading this while reclining in a bubbling spa pool, then there could be more to worry about than the pages getting soggy. The water spray could harbour harmful bacteria.

Britain's first case was 72-year-old Jean Winfrey, who has a spa in her house near Peterborough, Cambridgeshire. Doctors found abnormalities in her lungs, possibly caused by an inflammation called sarcoidosis, thought to be an allergic response to infection.

Medical experts in Britain were apparently baffled: the British Lung Foundation had never heard of the condition and specialists have urged doctors to take spa use into account.

Things are different in America, where hot tubs have been popular for years. The US Centre for Disease Control has tracked respiratory infections associated with spas and pools for some time and warns against placing them indoors.

Technically, it says, hot tub lung covers several conditions, including pneumonia and a hypersensitivity reaction that is not an infection. The most common culprits are mycobacteria. They live in the slime that forms on the inside of wet hot tub pipes, until the high-speed water jets break them off and fling them into the air among the bubbles.

Otis Rickman, a pulmonary specialist at the Mayo clinic in Rochester, Minnesota, says that many people do not clean their tubs and change the water often enough. The warm temperature makes matters worse: chlorine loses much of its disinfecting power, while the bugs thrive.

Why are sun creams getting stronger?

It is finally dawning on us that sitting out in the sun is not such a good idea, hence the number of people now applying cream as strong as factor 60.

'The whole deal with sunscreen is that one wants to minimise the amount of ultraviolet radiation that gets through to the skin's surface,' says Richard Groves, a dermatologist at Imperial College London. 'We know that UV radiation is of high energy, and is responsible for skin cancer formation.'

The sun protection factor (SPF) of a sunscreen does what it says on the tin. 'If it normally takes you, in the midday sun, 10 minutes to burn and you put on a factor two sunblock, it will then take you 20 minutes to burn,' explains Groves.

There are two type of sunscreen on the market: those that physically reflect the harmful ultraviolet light and those that absorb it.

The first are typically thick white pastes as seen on the noses of Australian cricketers, although more recent versions are transparent. Their active ingredient is zinc or titanium oxide. 'The nice thing about those reflective sun blocks is that they tend to have a very broad spectrum of activity,' says Groves. In other words they will block out UVA and UVB, the two types of UV light. The former causes wrinkles and the latter is the main cause of skin cancer.

The second, absorbing sunscreen contains chemicals such as oxybenzone, salicylates and cinnamates. These are transparent and are more active in the UVB part of the spectrum.

'We know that the risk of developing skin cancer is directly proportional to the amount of UV light that gets to the skin – primarily UVB,' says Groves. There is evidence that the closer you live to the equator, the more likely you are to develop skin cancer. And judging by the increasing uptake of high SPF sunscreens, it seems that people are taking the risk seriously.

The skin's own mechanism for protection is remarkably clever. Cells in the epidermis called melanocytes inject melanin (the skin's protection against UV light) into the surrounding cells, which then use the chemical to protect their nuclei. 'That's where the DNA is and it's the DNA that gets damaged by ultraviolet light,' says Groves. 'If you cut a section through these cells, you can see a little hat of melanin sitting over the nucleus. It shows you how sophisticated the protection mechanism is – it's not just randomly distributed through these cells.'

But clearly that's not enough for those of us who want to get a tan. The incidence of melanoma, the 'killer' skin cancer, is about 5–8 per 100,000 population per year in northern Europe. This may seem low but it is a disease of young adults and, if not caught early, is almost impossible to treat. So remember, sun-seekers, experts recommend that the lowest SPF you should use is 15.

What does it take to cut off your own arm?

More than dedication, that's for sure, as Aron Ralston found out. Ralston, from Aspen, Colorado, was climbing alone in a remote canyon 150 miles south-east of Salt Lake City when he dislodged an 800 lb boulder that fell on his right forearm, pinning it, and so him, against the rock face.

After three days and no sign of rescue, his water ran out. Two days later, Ralston figured it was decision time: lose the limb or lose his life. He pulled a tourniquet tight around his right bicep, got out a small penknife and set about hacking through his forearm. Having cut his arm back to a stump, he used ropes to lower himself 25 m to the bottom of the canyon where he set about finding help.

'People can do superhuman things under pressure, but this is different,' says Simon Lambert, an upper limb surgeon at the Royal National Orthopaedic Hospital in Stanmore. 'This is a conscious decision – I'm going to cut my arm off. It's not about a superhuman physical effort, this was an extraordinarily mentally strong man.'

Because he was climbing, Ralston may not have felt the full brunt of the pain when the boulder first landed on him. When you are doing something dangerous, natural endorphins start rushing around your bloodstream, ready to act as painkillers should the worst happen. But they would only have been effective for a few hours. By then, most of the crushed limb would have gone numb for good. 'After an hour and a half of an 800 lb boulder lying on your arm, the nerves and muscle tissue below the crush line will be dead,' says Lambert. 'The boulder would have done the major part of the amputation itself.'

Even if you can summon the determination to amputate one of your own limbs, there are a hundred ways to stuff it up. 'The simplest way to do it would be to go across the joint. Then you don't have to cut through the bone,' says Lambert. Ralston's arm may have been so fractured, he didn't need to cut through any bone. If so, it would have been a blessing of a kind. 'He was a climber and a skier and his bones would have been incredibly hard,' says Lambert.

Next, you need to work out how to tackle the arteries. Start chopping at the elbow and you have only one artery to cut through, but this splits below the elbow, into two that run down the forearm. Whether you cut through one big artery or two smaller ones makes little difference. If your tourniquet is not tight enough, you are in trouble.

Next for the chop are the nerves. There are four major nerves in the forearm, and they can be tricky to cut properly. Done well, the severed nerves retract back into the muscle. 'It is vital to get this right,' says Lambert. 'Otherwise you get an exquisitely painful lump on the nerve called a neuroma and you don't want one of those near your stump.'

It's not so hard if you are chopping off someone else's limb in an operating theatre. Lambert says that these days an amputation takes about 45 minutes to do properly with the right kind of scalpel and saw. It's positively sluggish compared to Astley Cooper, the most popular surgeon in London during the early 19th century. Cooper's amputations might not have been as precise as today's best efforts, but he reportedly often completed the job inside 15 minutes.

What strikes Lambert is the psychological strength Ralston showed. 'Doing the amputation yourself takes a special sort of person. That he was so clear-thinking and so determined to live, that he was willing to do that speaks volumes for his character. The question, am I going to do it myself, would tax even the most outrageously brave man,' he says.

Ralston joins an elite list of self-amputees. A few years ago, Doug Goodale, a lobster fisherman from Maine, used a knife to cut his arm off at the elbow after getting it caught in a winch. And in 1993, fisherman Bill Jeracki used a bait knife to cut his leg off at the knee after getting it trapped under a boulder while fishing in a stream in Colorado.

Just one last thing: don't try this at home, kids. It's very, very dangerous to do.

Babies & Children

Why are sperm counts falling?

Although many environmental factors have been blamed in the past, it's still a mystery.

Researchers from Aberdeen found there was a 29% drop in the average sperm count of more than 7,500 men who had attended a fertility clinic between 1989 and 2002. The researchers were quick to point out that the study might not be typical of the whole population, but it once again sparked fears of ever-decreasing male fertility and is not so far from the truth.

In the mid-1990s, researchers across Europe did find a steady decline in sperm quality over the previous 20 years. There was no decline in America or Finland in parallel studies. Many people blame the increased number of synthetic chemicals in the environment for falling sperm counts. Chemicals such as bisphenol A and drugs such as the Pill (which both act like the hormone oestrogen) are usually cited.

Michael Joffe, a reproductive epidemiologist at Imperial

College London, disagrees. 'The idea that it's weak oestrogens just cannot explain the observations. We know . . . that, if there's an oestrogenic effect, you need huge amounts of very potent oestrogen.' He says it is more likely to be chemicals that act as anti-androgens. These either interfere with the production or actions of androgens (hormones such as testosterone that control the development and maintenance of masculine characteristics).

The most common of this class of chemicals is DDE, the stable breakdown product in the body of DDT. 'There's been a lot of this stuff around in the 20th century,' says Joffe. But he says even this may not be the final answer. 'My guess would be that it's more likely to be something in the food than something in the air or water,' he says.

Whatever environmental factors cause the declining sperm count, the consequences are far-reaching – scientists agree that there is probably some genetic damage involved, too. So any problems the father has are passed to his son.

Does age affect a man's fertility?

For years, the impressive case of Charlie Chaplin, who fathered a child in his late seventies, has been the lecture room example of man's lifelong fertility.

While studies have nailed down a significant drop in female fertility from the age of 35, unravelling the effect of age on male fertility is a tougher nut to crack. 'It's been extremely difficult because the older men we need to study tend to be married to older women who have gone through the menopause,' says Allan Pacey, a senior lecturer in andrology at Sheffield University.

Men produce sperm for the duration of their lives and indicators such as sperm count and swimming ability change very little with age, which could suggest that male fertility might not dip either. But that is not the case. In 2000, researchers in Bristol looked at the rate of successful pregnancies among thousands of couples. They found men aged 40 and older were half as likely to get their partners pregnant as men under 25.

'What they couldn't tell was whether it was a biological effect, was it down to the fact that older men simply have less sex, which we know to be true, or that the longer people have been together, the less sex they have,' says Dr Pacey.

What is known is that as men age, the DNA in their sperm accumulates damage. 'If the DNA is damaged, they might get fertilisation, but the embryo won't develop or it will miscarry,' says Dr Pacey. Researchers in California reported a study of 70,000 couples which showed that men aged 50 and above were more than four times more likely to have a child with Down's syndrome. Other studies have shown an increased prevalence of schizophrenia.

'Much of the time, we don't see the effects of older paternal age . . . The only time we can see what happens is when you get old, rich and powerful guys who somehow have much younger wives,' says Dr Pacey.

Can babies survive abortions?

It's very unusual, very distressing, but very rare. Only babies aborted after about 22 weeks are developed enough to stand a chance when medically induced, and then only if the doctor performing the termination has failed to give them an injection to stop the heart first.

Hardly any of these babies manage to cling to life for more than a few minutes: the handful of aborted babies surviving beyond that are usually from pregnancies ended for reasons of very severe abnormality beyond the 24-week legal limit.

Stuart Campbell, who worked as an obstetrician at St George's Hospital in London and now offers high-tech 3D scanning at a private clinic, says some consultants aren't skilled enough to administer the lethal potassium chloride injection. Even then, most aborted foetuses that are brought out severely premature cannot use their lungs and die as soon as they leave the womb – but there are some who don't.

'Many of them do show signs of movement,' Campbell says. 'There is reflex activity in the limbs, the heartbeat will be there and they may actually make feeble attempts to breathe. It won't be successful and will only last a few minutes, but it's extremely distressing and it's the reason why I always insisted on stopping the heart before the procedure is done.'

Foetuses as young as about 16 weeks can make limited attempts to move when they are induced. 'It's wholly unacceptable to have a baby born and show even feeble attempts to survive,' says Campbell.

Before a foetus reaches 14 weeks it is small enough to

remove by using a suction tube without inducing labour. After that the situation becomes more complicated and a variety of techniques can be used. Most doctors now use drugs to stimulate contractions in the uterus walls to push out the baby; other procedures involve dilation and forceps.

Can we all have 'designer babies'?

Depending on how you define the term, designer babies are either already here, nearly here, or still remain in the realm of science fiction.

In Britain, the debate is over 'saviour siblings' – embryos selected on the grounds of being good tissue donors to help cure ill siblings.

So technology already provides some people with a choice over the genetic makeup of their newborns. Women undergoing fertility treatment can have embryos screened for a range of genes linked to different conditions. In Britain, it is also legal for embryos to be selected for gender, but only if there is a good medical reason. For example, parents who know that their child may inherit haemophilia can have embryos screened for sex because while it directly affects males, females are simply carriers of the condition.

In some countries, such as the US, India and certain states in Australia, parents can choose the sex of their child for what is often termed 'family balancing'. Tests for eye colour and hair colour might also be technically possible.

The process itself is horrendous. First the woman needs hormonal treatment to stimulate egg production, and only some of these will become viable fertilised embryos. Then, for genetic analysis, a few cells have to be taken from the embryo – which is unlikely to be risk-free – to check its genetic makeup. Then, any eggs that pass the test have only a 25% chance of being carried to term. 'To think someone would go through that just to ensure their baby has blue eyes or blond hair is highly unrealistic,' says Mary Petrou at University College Hospital perinatal centre in London.

The prospect of tinkering with the genes of an embryo to ensure that the child becomes a musical genius or a sporting legend is science fiction. Even if genes were found that predisposed a person to excel in a particular skill, the environment he or she grew up in would contribute hugely.

One way that babies can be 'designed', up to a point at least, is by carefully choosing the donor. Fertility clinics provide information on donors, such as body shape, hair and eye colour, which may influence a child's appearance. 'It takes you as far as going down the pub and picking somebody you like the look of,' says Stuart Lavery, an expert in pre-implantation genetic diagnosis at Hammersmith Hospital in London.

Are excessive ultrasound scans bad for unborn babies?

Probably not, but we can't be certain. 'Ultrasound is a [source of] energy. At the extreme it will cause problems,' says David Liu, head of the foetal care unit at the City Hospital in Nottingham. 'At the low extreme it doesn't cause anything at all. It's just like talking. The [ultrasound units] designed for scanning are designed in such a way that it does not cause any harm.'

Some doctors have raised concerns that some parents were exposing their unborn babies to too many scans. They warned that the increasing practice of making DVD movies of unborn babies or having 3D pictures taken might carry unknown risks.

Under normal NHS procedures, two brief scans are taken to check for abnormalities. But private clinics have developed a lucrative sideline in non-medical scans for parents to keep mementos of their child before birth.

Liu says the concerns are unfounded. 'When there's some abnormality or we're not clear about the structural defect of a baby, we sometimes spend hours scanning them,' he says. Scanning for a video or a picture would be for minuscule amounts of time by comparison. In addition, says Liu, ultrasound scans have been used for around 20 years with no ill effects.

Previous research on potential risks has suggested that excessive scanning could cause growth problems. But Liu disputes the findings. He says the study involved unusually small babies that were scanned more often anyway.

He says that perhaps those warning against non-medical scans are railing against the commercialisation of a medical procedure. 'Some of my colleagues are, likewise, unhappy that people are capitalising on a scan. It niggles them a bit.'

Do pregnant women make the best wine tasters?

Yes, if the bevy of wine buffs at Tesco's HQ in Hertfordshire are to be believed. When four of the supermarket's team of tasters became pregnant, all claimed their tastebuds became hypersensitive.

Scientific evidence backing up the tasters' claims is hard to come by, but smell expert Tim Jacob at Cardiff University says hormonal changes are known to affect one's sense of smell. And since taste is 75% smell, being pregnant could well make flavours stand out more. 'It's well known that a woman's sense of smell changes during the menstrual cycle and peaks at ovulation,' he says.

A heightened sense of smell could be an advantage for pregnant women, Jacob adds. Smell is crucial to mother–infant bonding in the first few days following birth. 'A mother can often tell her own child merely by its smell after a few days,' he says.

The problem with studying the effect of pregnancy on taste is that women are affected in a variety of ways. 'Not all women experience a change in their taste and even those that do can experience very different changes,' says Jacob.

Is it bad to opt for a caesarean?

From a purely medical perspective, there's little wrong with it. In fact, most obstetricians argue that birth by caesarean section is probably the safest thing for a baby. But that doesn't mean it's the best option for the mother.

The rate of caesarean sections has more than doubled in the past two decades. Nicholas Fisk, a professor of obstetrics and gynaecology at Imperial College London, attributes the rise to our increasingly risk-intolerant society. There are now twice as many women over the age of 35 having children as there were in 1989 and Fisk argues that older women find labour more stressful and can suffer more complications and are, therefore, more likely to request a caesarean section.

If complications arise in natural birth, a caesarean is the only option. 'If you're over 35 or you've got twins or your baby is a bit small or you're a bit short, or there are other problems, you've got a much higher chance of ending up with an emergency ceasarean,' he adds.

Once an operation has started there are three areas where doctors are particularly alert to complications – excessive bleeding, blood clots in the woman's legs and infection.

For mothers, there is definitely an increased risk of death with caesareans. Figures from the Royal College of Obstetricians showed that for two million births between 1997 and 1999, 69 women died, 40 of whom had had caesareans.

But the figures mask the fact that many operations are carried out under extreme situations. 'An emergency caesarean section is more dangerous because it is done when the woman is exhausted, is more likely to be infected and all the tissues are stretched,' says Fisk. 'And it's often in the middle of the night in a hurry.'

Modern techniques, however, mean that the problems are easier to manage. 'They're all done under epidural, all women are given antibiotics and most are given drugs to thin the blood a bit to stop any clotting risk,' says Fisk. He adds that there is a lot of work going into minimising the surgical aspects of the procedure to make it even less stressful.

After the operation, the patient can be incapacitated for up to a week and the pain of such a serious abdominal operation will cause them considerable discomfort for some time. Some groups have even suggested that the bonding between mother and baby can be adversely affected, particularly if the operation is done under general anaesthetic, but there is little data to support this.

Fisk argues that the practice of repeatedly electing for a caesarean could be damaging for some women. 'It's not a great idea to have a lot of caesarean sections. The moment you get into three and, particularly four, five and six, you expose yourself to a low but substantially increased risk of the placenta getting stuck over the scar.' This can be catastrophic for the baby.

What are the chances of survival for very premature babies?

More and more tiny babies born at less than 25 weeks in the womb are surviving – a triumph for medical science which also fuels the debate about the 24-week limit for abortion.

But Baroness Warnock, who charted Britain's path into embryo research and fertility treatment, believes there should be a cut-off point: a gestational age below which doctors would not try to keep every newborn alive. 'Some doctors and nurses get competitive about the triumph of keeping these tiny, premature babies alive,' she said.

In Holland, babies are not given medical treatment before 25 weeks. In the UK, not every hospital will pull out all the stops to keep a baby of less than 23 weeks alive. But it is a difficult decision, because the parents' wishes may run counter to the instincts of the medical staff.

That was the reason for the EPICure study in 1995, which was designed to provide clear evidence about the outcomes for very premature babies. The researchers studied babies born before 25 weeks and six days in the UK and Ireland, and attempted to follow up the survivors. The first thing they found was that most babies died. Of more than 1,200 born at 25 weeks or less, only 308 survived to six years. The risk of dying varied dramatically within a few weeks of gestation: at 22 weeks, 98% die, but at 26 weeks, 80% survive.

Of those 308 children, the researchers were able to follow up 241 at the age of six. Most were in mainstream education, but 37 were in special needs units. Four children were blind and two could see only light. Seven had profound hearing loss and seven more needed hearing aids. About 22% had cerebral palsy. Only the minority were severely disabled in some way, but 41% had moderate to severe learning difficulties.

How heavy can a baby get?

Something that might have crossed the minds of surgeons who delivered, by caesarean section, a 13 lb 13 oz baby.

'Mighty' Joe Griffin, as mother Sara dubbed the baby, may

be what statisticians call an 'outlier', but he's by no means the heaviest newborn. According to *Records and Curiosities in Obstetrics and Gynaecology*, by I.L.C. Ferguson, the heaviest normal baby born in recent times weighed 24 lb 4 oz in 1961 in Ceyhan, Turkey. That compares with the British median birthweight of just under 7 lb 7 oz.

Patrick O'Brien, a consultant obstetrician at University College Hospital in London, says that no one has found a mechanism in pregnant women that prevents babies growing beyond a certain weight. Instead, labour is triggered when the baby reaches a certain stage of maturity.

Babies are getting heavier, and a study by Pamela Surkan at the Karolinska Institute in Sweden explains at least some of the reasons. Her study showed that the number of heavy babies – over about 9 lb 14 oz – had risen from 3.71% to 4.6% between 1992 and 2001. During the same period, obesity among women rose to 36%, and smoking, which limits foetal growth by cutting oxygen in the placenta, fell from 23% to 11%. So better nutrition, and in some cases, too many calories, coupled with lower smoking rates explain why babies are getting heavier.

An increase in diabetes is also leading to heavier babies, says Karl Murphy, a consultant gynaecologist at St Mary's in London. Diabetic mothers have raised blood sugar levels, and since blood crosses the placenta, babies react by over-producing insulin and growth hormones, making them develop faster.

Does crying damage babies' brains?

If it does, then there must be a lot of brain-damaged babies around. 'If you think of the amount of crying that babies do, you would think biology would ensure it doesn't cause brain damage,' says crying expert Dieter Wolke, the scientific director of the Jacobs Foundation at Zurich University. 'I can't see how it'd happen.'

However, Margot Sunderland, who runs a conference and lecture organisation called the Centre for Child Mental Health in London, said that stress levels in babies who aren't comforted when they cry can get high enough, and remain high for so long, that it causes brain cells to die. That in turn can lead to neurosis and emotional disorders later in life, she says.

Researchers say that while animals exposed to very high levels of stress for prolonged periods can develop changes in their brain structure, stress from crying has never been shown to cause such damage. 'If it were true, it'd be a surprise,' says Annette Karmiloff-Smith, professor of neurocognitive development at the Institute of Child Health.

Crying is an immensely useful mechanism for raising the alarm that all is not well, says Wolke. Before the age of six months, crying is almost always a genuine plea for help, rather than simply a way of grabbing attention, he says. On average, a baby cries for two-and-a-half hours a day for the first three months of its life, but for only about an hour once it is one year old.

Wolke says that once an infant is older than six months, it is safe to start leaving the baby to cry for longer periods if they are doing it purely to get attention – so-called controlled crying.

'You should never leave a baby to cry until they are about six months old, because before that age, they don't have the ability to cry just to get attention, they are doing it for a reason,' he says.

How many vaccinations is it safe for a baby to have at once?

Believe it or not, a baby has the theoretical capacity to tolerate 10,000 vaccines at any one time. Vaccines work by introducing the body to various antigens – foreign objects that stimulate it to make antibodies, which then fight them. Antigens can come in many forms: bacteria, viruses, proteins, toxins or even transplanted organs. Vaccines normally contain some form of the disease being protected against.

Whether or not multiple vaccines overwhelm an infant's immune system was the subject of a study published in the journal *Paediatrics* in January 2002 by Paul Offit of the Children's Hospital of Philadelphia. He came to the 10,000 vaccine figure by first working out how much antibody would be required to fight a particular antigen.

Assuming that there are about 100 antigens in every vaccine, and calculating the amount of time it would take a baby's immune system to manufacture enough antibodies for each, Offit was able to work out the baby's theoretical capacity.

Opponents are not convinced. 'The reasons we've always

had concerns over multiple vaccines is . . . that safety studies in the UK have never been long enough or effective enough,' says Jackie Fletcher, a spokeswoman for the parent support group Jabs. 'With the single vaccines, because they have been around and in popular use for 40 years, we've got safety track records that are proven.'

But Helen Bradford, a researcher at the Institute of Child Health, says the new vaccine is much safer. 'This is actually a really good development,' she says. 'From the minute they're born, [babies] are exposed to a huge number of antigens every time they breathe,' she says. 'What would happen to a baby if their immune system was being overloaded? One of the possible things would be that they got ill after vaccines, they got infections. There have been a number of studies where they've looked at hospitalisation for infections following vaccinations and found that vaccinated babies have fewer infections after they are vaccinated than unvaccinated babies.'

The number of antigens a baby is exposed to via vaccines has decreased during the past century, despite the increase in the actual number of vaccines being used. In 1960, babies routinely received five vaccines containing more than 3,000 types of antigen: now they receive about 11 vaccines with up to 126 antigens. By Offit's calculation, these 11 vaccines 'use up' about 0.1% of the baby's immune system.

Why is breast best?

Breast-fed babies get exactly the right proteins, fats, carbo-hydrates, vitamins and minerals they need in exactly the right amounts as well as increased protection against infection. And there is evidence to suggest that the benefits they receive early on could last for life – the World Health Organisation recommends that breast-feeding as the best form of nutrition for the first six months of a baby's life.

'Everything that's in breast milk is very well absorbed by the infant,' says Toni Steer, a nutritionist at Cambridge University.

The protein in breast milk is predominantly whey and that is easily digested. In contrast, the protein in infant formulas tends to be based on casein, which is found in cow's milk.

Human milk's fat is rich in long-chain polyunsaturated fatty acids. 'We think that they are advantageous because they are involved in the development of things like neural cell membranes,' says Steer. These are important for the development of the brain and nervous system and there could be a link between these and cognitive and mental development.

The fats also play a role in the development of arteries and veins. Studies on six-year-old children found that those who had been breast-fed as babies had lower blood pressure. Since blood pressure tends to track from childhood to adult life, and high blood pressure is the risk factor for things like heart disease, getting the polyunsarated fats in early is a definite health bonus.

According to Steer, formula milk does not contain high quantities of these fats.

Lactose provides nearly half of the energy value of breast milk as well as promoting the absorption of calcium and other trace elements. With the aid of oligosaccharides – another breast milk ingredient – lactose helps to develop the baby's gut.

When a baby is born, he or she is germ free. 'If an infant is breast-fed, what we tend to see is that they develop gut micro-flora, which is considered beneficial in reducing the incidence of things like gastrointestinal infections,' Steer says.

A mother's own immunity to the things in her environment is also passed on, to some extent, through her milk. Antibodies such as immunoglobulin, for example, help the baby fight infections.

Despite the great benefits conferred by breast milk, Steer says that there is nothing wrong at all with formula milk in a nutritional sense.

'The research that goes into the development of formula milk is enormous and they have consistently tried to produce formula milk which matches breast milk as closely as possible,' she says.

But there still seem to be certain things that researchers just cannot mimic – the growth hormones and digestive enzymes, for example. Maybe they'll get there one day but until then, there's no doubt at all which is best.

What causes cot deaths?

Nobody knows. Sudden infant death syndrome (SIDS) is the leading cause of death in babies over one month old. 'It takes more lives than meningitis or leukaemia or any form of cancer,' says Joyce Epstein, director of the Foundation for the Study of Infant Deaths. 'More babies die of cot death than are born with spina bifida or any range of illnesses. The peak age is 2–4 months. It can happen in much older babies, even over a year, but the cot-death rate drops off dramatically after 6 months.' In 2002, 342 babies died in unexplained circumstances.

Finding out why is not simple. 'In terms of research, it's a difficult thing to do because you don't know which babies are going to die,' says Lucilla Poston, a foetal health researcher at King's College London. 'So it has to be epidemiological studies that do populations and correlations with lifestyle and environment.'

Research has led to successes in reducing cot deaths. 'Until about 1989, there were about 2,000 deaths a year. Then in 1991, we were able to launch a campaign advising parents that it's safer to sleep babies on the back than on the front,' says Epstein. 'Within one year, there was a 50% drop in cot death. That doesn't say anything about the causes of the death, of course, that just talked about risk factors.'

There is also research on why SIDS happens in the first place, which focuses on areas including respiratory and cardiac systems, genetics, immunology and toxicology.

Genetics is a very promising avenue in trying to warn some parents that their child might be at risk. 'There is no gene associated with cot death, but people are now looking at family histories and seeing if they can identify families at risk,' says Poston.

Is it safe for children to take Ritalin?

If not, then we're storing up a whole heap of trouble. Figures show that use of the drug given to calm hyperactive children has soared 100-fold in Britain in the past decade.

Ritalin has been approved for use in children over five years old who suffer from attention deficit hyperactivity disorder (ADHD), but many doctors prescribe the drug – also known as methylphenidate – for children as young as 18 months.

The drug is widely considered safe, but few long-term studies have been done because of the ethical difficulties of experimenting with children. And some scientists are concerned how the drug – a powerful stimulant shown to have a similar effect to cocaine – could affect developing brains. One high-profile investigation in 2001 did suggest that Ritalin triggered changes to brain function. But Joan Baizer, who led the research at Buffalo University, New York, says its results were largely exaggerated. 'It was a preliminary report on basic research in rats ... it doesn't imply anything about a long-term health effect,' she says.

Dave Woodhouse, who runs an ADHD clinic at the University of Teesside, says that some evidence is emerging that children taking Ritalin for several years show reduced cognitive ability, but this work is also at an early stage and has not yet been published.

Food & Drink

Are cloned animals safe to eat?

We can't be absolutely sure, but the key conclusion scientists have reached is that cloned sheep, pigs and cows should be as safe to eat as their non-clone counterparts. And in our post-BSE, intensively farmed world, take that whichever way you like.

The suitability for the barbecue or bacon butties of animals created in a laboratory as identical copies of their parent is still largely an academic discussion, because no cloned animal is thought to have entered the food chain.

It's also currently far too expensive, at about $20,000 (£12,000) a clone, for the companies involved to even consider selling their prized animals as burgers.

'People are not producing clones for that purpose,' says Scott Davis, president of the US animal biotechnology company Viagen. Davis says that it is probably only the offspring of cloned animals that would be sold for meat.

Cloning could make sense for farmers as, in theory, it would allow them to easily produce lots of animals from a single prized specimen. A handful of American companies including Viagen are already producing cloned farm animals (mainly cows) for agriculture rather than scientific research. This prompted the US Food and Drug Administration to investigate the possible health risks, and it said that 'food products derived from animal clones and their offspring are likely to be as safe to eat as food from their non-clone counterparts'.

The defects and abnormalities that cloned animals can suffer have been well publicised since Dolly the sheep was created in 1996. Clones can suffer from abnormal growth, obesity and premature ageing, while Dolly herself was destroyed after developing arthritis and a lung disease – unusual but not unheard of in a sheep that age.

Biotechnology companies developing cloned animals complain that these problems have been exaggerated as part of the campaign against human cloning, but Ian Wilmut, the geneticist who led the team that created Dolly, insists that they need to be considered before cloned meat is approved.

'I think it is extraordinarily unlikely that cloning would change an animal in such a way that food from it would be unhealthy to anyone who ate it, to me the greater issues are concerned with the welfare implications for the animals,' Wilmut says. 'The experience still is that there are considerable problems at birth and in some cases after birth for cloned animals.'

A spokesperson for the UK's Food Standards Agency says that meat from cloned animals would be classed as a novel food, so companies trying to sell it in Europe would require a special licence.

Is organic food any healthier than conventional food?

It depends who you talk to. According to the UK Food Standards Agency (FSA) there are no health benefits to be gained from eating organic food. 'In our view the current scientific evidence does not show that organic food is any safer or more nutritious than conventionally produced food,' says Sir John Krebs, chairman of the FSA.

However, a number of studies have shown a clear difference between organic and conventionally farmed food.

Scientists in the US conducted a study into pesticide residues on organically grown plants. The results clearly showed that conventionally grown crops were six times as likely to be coated in pesticide residues. As yet there is no clear evidence to show what impact pesticides might have on human health, but many scientists are concerned about them. 'Organic food clearly offers consumers the best choice to avoid pesticides in their diets,' says Brian Baker from the Organic Materials Review Institute in Oregon and lead author on the study.

Campaigners claim eating organic also means that you are likely to be getting more vitamins and minerals per mouthful, especially vitamin C, magnesium and iron. 'A US study in 2003 found that organic crops had higher average levels of all 21 nutrients analysed,' says Gundula Azezz, policy manager from the Soil Association.

Studies have shown that organic milk and meat contain higher levels of the essential fatty acids, such as omega 3 and conjugated linoleic acid (CLA). These are thought to be necessary for metabolism and can help to prevent many medical problems.

'It is difficult to tease out the evidence to show that organic food is healthier, but eating organic removes the uncertainty associated with eating unnatural products,' says Azezz.

Does the GI diet work?

GI, or glycaemic index, is a measure of how quickly different foods raise blood sugar levels. Unsurprisingly, sugary foods such as cakes and biscuits have high GIs (and for the sake of the diet are colour-coded red), while less celebrated treats such as lentils and porridge have low GIs. These are colour-coded green, hence the traffic lights.

Follow the GI diet and you will cut down on the high-GI foods and tuck into more 'slow-burning' pulses, bread made with unrefined flour, and fruit and vegetables.

The diet was originally used to help diabetics control their blood sugar levels, but in practice it is little more than what nutritionists recommend as healthy eating.

'It's the advice we've been giving for the past 10 years, but with these added labels of high and low GI,' says Claire MacEvilly, a nutritionist at the MRC Human Nutrition Research Centre in Cambridge.

The jury is still out on whether the GI diet works. Some large studies have shown that those on low-GI diets tend to have smaller waistlines and are at a lower risk of heart disease and type II diabetes, but studies claiming the diet leads to impressive weight losses have been greeted with caution.

Advocates of the diet say it works because low GI foods make people feel full for longer, and so banish hunger pangs. But evidence is scarce.

If the diet works at all, it might just as likely be down to low-GI foods containing fewer calories.

Why does Tony Blair eat black bananas?

Mr Blair, according to reports, eats two black bananas at lunchtime in the belief that some ingredient in the overripe fruit helps boost his energy levels, keeping him going on just five hours' sleep a night.

But as an energy-boosting strategy, the black banana diet is greeted with bemusement by experts. 'I don't know of any magical properties of black bananas that would really boost your energy levels after a lack of sleep,' says Amanda Johnson, of the British Dietetic Association.

Bananas, whether they be yellow or black, are undoubtedly a good source of energy if you need staying power. Sugary foods may produce a hard and fast energy rush, but it is always followed by a dip that can leave you feeling sluggish. Bananas are digested more slowly and release their energy steadily.

Bananas are high in starch, but as they ripen, the carbohydrate is broken down by enzymes. At the same time, the amount of sugar in the fruit increases, explaining why ripe fruit can taste sweeter. Although these changes are measurable, they would have at most a negligible effect on how energy is released from the fruit when eaten. 'When you look at the whole context of physiology and other foods you eat, it won't have a noticeable impact on digestibility or energy release,' says Johnson. Scientists at the Medical Research Council's Human Nutrition Research Centre in Cambridge agreed, adding that little research has been carried out on the nutritional content of very ripe fruit.

Is it safe to drink tap water with added fluoride?

Common sense says yes, because millions of people around the world have been swigging it back for decades with few obvious problems. Although a handful of studies have linked artificial fluoridation to a range of conditions including cancer, most experts believe the benefits significantly outweigh the risks.

About 10% of Britons drink fluoridated water, mainly in and around Birmingham and Newcastle, where water companies add trace amounts to the mains supply. Others live in areas where the water is naturally high in fluoride.

Fluoride promotes a chemical reaction in tooth enamel that strengthens it, and draws replacement minerals, including calcium, to fill holes. This means levels of childhood tooth decay in Birmingham are roughly half those in Manchester. But the same chemical reaction can also affect bones, hence concerns that increased doses of fluoride could promote conditions such as osteosarcoma, a rare bone cancer. And raised levels of fluoride are known to cause dental fluorosis –

unsightly mottled teeth. But after reviewing the evidence, two separate British studies recently gave fluoride the all clear.

'Routine public health monitoring is right, but there is ample evidence for, say, Manchester health authority, to fluoridate its water supply,' says Professor Mike Lennon of the British Fluoridation Society.

The British Dental Association says a targeted approach is needed, focusing on deprived areas where tooth decay is most common. Areas such as the home counties don't need it, Lennon says.

Can you eat junk food and stay healthy?

No. In his docu-film, *Super Size Me*, Morgan Spurlock spent 30 days eating McDonald's meals for breakfast, lunch and dinner. As he piled on the pounds, doctors monitored his soaring level of body fat. His libido also diminished and he started to feel depressed and tired. By day 22, medical specialists warned him he was seriously damaging his health and one doctor even said his liver is turning to 'paté'. So what is it that is so bad about burgers and fries?

'It is very difficult to say how often it is safe to eat junk food,' says Susan Jebb of the MRC Human Nutrition Research Centre in Cambridge. 'One chocolate bar won't kill you, but 10 a day is obviously not going to do you any good.' The problem seems to be that junk foods provide lots of energy, but not many nutrients. 'A junk food meal provides such a high proportion of your daily energy needs that it is hard to find food that is nutrient-rich enough to make up the rest of your daily diet, without giving you too many calories,' says Webb.

Spurlock's extreme diet brought him close to a serious liver condition known as non-alcoholic steato hepatitis. 'A healthy liver contains virtually no fat, but his high-fat diet will have led to fat infiltrating the liver,' says Jebb. 'The effect will have been similar to force-feeding a goose to make foie gras and the liver will have started to have visible blobs of fat in it.'

As well as gaining over 7 kg in weight, Spurlock also experienced an increase in two types of body fat, cholesterol and triglyceride. 'Cholesterols come from saturated fats, while triglycerides come from refined carbohydrates like burger buns and milkshakes,' says Jebb. Both of these fats are known to increase the risk of heart disease and diabetes.

Most nutrition experts don't have a good word to say about junk food. Loranne Agius at the Diabetes Research Centre at Newcastle University says: 'It is hard to justify a single advantage from eating junk food. It is only worth eating if the alternative is starvation.'

Is it safe to eat mud?

Known to the experts as geophagia, mud meals can be traced back to ancient Greece. 'Certain foods they ate were actually toxic, but by eating clay, the clay absorbs the toxins from the food, so it didn't cause them any harm,' says Barry Smith, a chemist at the British Geological Survey and one of the world's few geophagia experts. Animals have also discovered the benefits of eating mud – notably birds in the rainforests of

South America which eat clay from riverbanks to absorb dangerous poisons in the food they eat.

Whether it's safe, or even beneficial, to eat soil is debatable, depending at least on your nutritional state and the type of soil you're tucking into. Geophagia is still fairly common in parts of Africa and some scientists argue that, like nail-biting, it's a habit to be discouraged. Just as clay can strip toxins out of poisonous food, so can it remove useful nutrients, potentially leading to malnourishment, or so the argument goes. But others say eating soil is simply satisfying the body's craving for certain minerals.

Studies of groups in Africa show that soil can help to replenish levels of magnesium and iron. 'After the rains, people will say that the smell of the soil is so intoxicating that they must eat it,' says Smith. 'If you take into account the amount of soil they're eating, which can be up to 100 g a day, it corresponds to roughly the recommended daily intake of iron and magnesium.'

As far as eating British soil goes, Smith says he would play safe and advise against it. Natural soils, which have never been contaminated by industrial pollution, can still contain low levels of arsenic, lead, cadmium, uranium and nickel, the combined effects of which are unknown. More polluted soils could also contain toxic organics including PCBs and dioxins. 'What scientists are still trying to get their heads around is what happens if you're exposed to low levels of all of these together,' he says.

Is chocolate good for you?

It's a perennial question revisited in research by scientists at Imperial College London, which claimed that the theobromine in chocolate is a third more effective than codeine at stopping persistent coughs.

So, to run through those health benefits: in 2001, a study published in the *Lancet* showed that chocolate contained significant amounts of antioxidants in the form of flavinoids – these prevent the build-up of coronary artery plaque. In particular, catechin (also found in tea) protects the immune system and the heart. It reinforced a study carried out two years earlier at the University of Scranton, which showed that flavinoids were even better than vitamin C in preventing the bad LDL cholesterol in the blood from being oxidised and damaging the heart. Chocolate also contains several trace minerals including manganese, potassium, magnesium, phosphorus and calcium.

But this is no licence to gorge. Chocolate contains lots of saturated fat (in the form of stearic acid) and sugar while fruits and vegetables contain more of the antioxidant flavinoids. In 1999, a study published in the *American Journal of Clinical Nutrition* showed that the risk of coronary heart disease thanks to the fat in chocolate was even more than the risk posed by other harmful fats. As the editorial in that issue concluded: 'Unfortunately, for chocolate lovers, chocolate's high content of stearic acid puts it in the same category of risk of coronary disease as meat and butter – i.e. pathogenic.' Therefore, even if theobromine proves to be the best way to treat persistent coughs in the future, it's more likely that your doctor will prescribe you a pill than a Mars bar.

The solution? Eat bitter dark chocolate, which contains less of the bad stuff but still has the flavinoids.

What is so good about probiotics?

The word first popped up to describe the microbes that turn fresh milk into yoghurt, kefir, sour cream, buttermilk and cheese. But probiotics are now part of a £135m food supplement industry.

Consider the human as a perambulating apartment block, permanently colonised by bacterial squatters. An adult consists of 100 trillion cells, but is home to 10 times as many microbes. They cling to the skin, and 80 species inhabit the mouth. At any time, there could be 500 species of microbe in the gut, each numbered in their billions. They are not all good guys, by any means, says Anne McCartney of Reading University. Two lines of research led to an interest in modern probiotics, she says: one was the observation that yoghurt-drinking Bulgarian peasantry seemed to have more than their fair share of lusty centenarians; the other was that breast-fed babies did better than formula-milk babies, and the difference in each case seemed to lie with *Lactobacilli* and *Bifidobacteria*.

Jeremy Hamilton-Miller, of the Royal Free Hospital in London, tested more than 30 probiotic products, and wants to see clearer labelling, better guidelines and more convincing scientific trials. 'A certain type of probiotic is effective under certain conditions,' he says. 'You can't just pick any one. There are thousands of different strains and only a few have been shown by clinical trials to have specific clinical effects.

'We don't know exactly how they work. We know that they stimulate the immune system. We know that several of them produce substances that are antagonistic towards specific pathogenic microbes. We also know that they produce compounds of short-chain fatty acids that are very good nourishment for the intestinal cells. We also know that they bind to receptor sites that otherwise may be available for pathogens. Put all those together and you have a fairly potent mix.'

Are barbecues seriously bad for your health?

Yes, if you eat too much. And we have also been warned that sausages sizzling in the garden could deliver a dangerous dose of dioxins.

The French environmental group Robin des Bois tested the fumes associated with the two-hour grilling of four steaks, four turkey cuts and eight sausages over a barbecue, and says that it counted far greater doses of polychlorinated compounds, known as dioxins, than would be permissible at the outlet of a commercial incinerator chimney. In fact, levels approached those associated with smoking 20,000 cigarettes.

This is not that surprising, in the light of previous studies on smoke from fires, and there is no denying dioxins, also a constituent of Agent Orange, are carcinogenic. Lab tests on animals have repeatedly confirmed this, and workers in the chemical industry exposed to high levels of dioxins are at increased risk of cancer, heart disease and diabetes. They are also linked with birth defects.

A dioxin-free world would be quite hard to achieve, however. They pour from rubbish-burning incinerators, garden bonfires, forest fires and the family fireplace. Dioxins also dissolve in fat, and tend to accumulate in beef, dairy products, pork, milk, chicken, fish and eggs.

We must also remember that for all prehistory and almost all recorded human history, humans cooked over open fires of wood, charcoal or dung. Most humans still do. In spite of this, human numbers have gone from 1 billion to more than 6 billion in the past 200 years.

So the message is: go easy on the charred fillet and sooty sausage, but also save that pinch of salt for the health warnings.

Is eating live insects a bad idea?

Not if you're a television executive, as demonstrated by the stellar ratings for *I'm a Celebrity Get Me Out of Here!* in which Z-list celebrities feast on a selection of live jungle creepy crawlies.

The show's bush tucker expert makes sure the reluctant diners chew their food to kill it. 'You don't want this stuff going down alive,' he said, which certainly made great TV, but was it strictly necessary?

'Doing this on a television show it's probably not a bad idea to get them to bite the stuff in two,' says Tom Turpin, an entomologist at Purdue University, Indiana. 'I could imagine a big cockroach swallowed in their death throes with their claws and things could create some trauma in the stomach lining, so I suspect dead is better.'

People owning certain lizards as pets, he adds, are advised not to feed them live mealworms – one of the jungle dishes – as they can burrow into and damage the lining of the digestive system.

Turpin says he is prone to a spot of entomophagy (eating insects) himself, mainly as part of demonstrations to students and the public. 'Each insect is different but as a package they probably bring a better nutrient balance than almost any other single thing you could eat,' he claims.

And the taste? 'Mealworms have a taste very closely associated with whatever they've been eating,' Turpin explains. 'But take them off their food for a few days so you can really taste the mealworm, and then frankly I can't describe it.'

Would eating a Stone Age diet make us healthier?

If only things were that simple. Supporters of the so-called Stone Age diet argue that farming practices introduced about 10,000 years ago are ultimately harmful to human health, and that if our hunter-gatherer ancestors evolved without eating dairy products or cereals then we shouldn't eat them either. Instead, they say, we should only eat plenty of lean meat and fish, with fruit and raw vegetables on the side.

The idea, also called the caveman, hunter-gatherer or paleolithic diet, has been around for decades and is regularly recycled.

According to Lauren Cordain, a nutritionist at Colorado State University who published a book about it called *The Paleo Diet*, those following the meat-dominated menu 'lose weight and get healthy by eating the food you were designed to eat'. He says there is increasing evidence that a Paleolithic diet can prevent and treat many common western diseases. Studies of islanders in Papua New Guinea who still live a hunter-gatherer lifestyle show they rarely suffer heart disease.

But other nutrionists argue that, as with the closely related Atkins diet, cutting out whole food groups such as cereals is just not a good idea.

'I would recommend anybody to eat lean meat and raw vegetables,' says Toni Steer of the MRC Human Nutrition Research Centre at Cambridge. 'But what you're asking people to do is cut out a food group for which we have a lot of evidence to show is good for your health.'

Archaeologists say it's not even clear exactly how much of the various foods people actually ate during the Stone Age (broadly defined as from two-and-a-half million years ago until 10,000 years ago).

'There was no one Stone Age diet – diets of the past varied greatly,' says John Gowlett, an archaeologist at Liverpool University. People in Africa probably ate less meat than many people think, he says, while those in the northern, icy regions were forced to eat only whatever animals they could catch.

'I'm not convinced that we know what Stone Age man ate,' agrees Andrew Millard, who researches ancient health and diet at Durham University. 'The evidence we have is heavily biased towards the meat component of the diet. We get bones from animals they have eaten but we don't get the remains of any vegetables they have eaten because they decay.'

Millard adds that there is good evidence that later Stone Age cultures in the Near East regularly collected and ate wild cereals and it's possible that the practice was more widespread.

What's the best way to seal wine?

According to John Corbet-Milward of the Wine and Spirit Association, the increase in global wine production and surge in consumer demand has placed cork manufacturers under enormous pressure. The result, he says, is that some bottles on the shelves of the supermarkets and off-licences have been sealed with poor-quality cork. 'What has been unacceptable to the consumer has been the high level of tainted or corked wines caused by poor-quality corks,' he says.

The problem has pushed some suppliers to use screwcaps. Although frowned upon by many wine buffs, screwcaps are certainly effective, says David Berry Green, a buyer with Berry Bros & Rudd. They avoid the problem of corked wine because the chemical responsible for the corked taste, trichloroanisole (TCA), is produced by fungus living on the cork. Ironically, the fungus converts chlorine – often used to rid the cork of contaminants – into TCA.

High-quality corks are less likely to give rise to corked wine because they are not as porous, so the fungus and TCA it produces are less likely to find their way through the cork into the wine. Properly cut, pores should run sideways through the cork so as not to provide a route from the outside world to the wine.

Berry Green says screwcaps are perfect for sealing wine, especially bottles to be consumed early. By forming a water-tight seal, they also prevent the wine from oxidising, another problem associated with low-quality, highly porous corks.

Will feeding prisoners vitamin supplements make them behave better?

It can't hurt. Eating a healthy diet is known to be good for general physical and mental well-being.

Catherine Collins, a clinical dietitian at St George's Hospital Medical School, says research on the connection between diet and behaviour is limited, with much of it done by alternative health organisations with interests in selling food supplements. The relationship is complex and as yet there are no definite explanations.

Even so, a study published in the US showed that children in Mauritius were significantly less likely to grow up with criminal records if fed an enriched diet.

The British research charity Natural Justice, which focuses on the causes of antisocial behaviour, brought together physiologists and psychologists to compare the behaviour of 231 young adult prisoners before and during a regime of extra vitamins, minerals and fatty acids. The study showed that those eating food supplements committed an average of a third fewer offences than those without the supplements.

'Antisocial behaviour in prisons, including violence, is

reduced by vitamins, minerals and essential fatty acids, with similar implications for those eating poor diets in the community,' the researchers concluded. As well as the dietary quality of meals, Collins says, behaviour can depend on everything from family dynamics to socio-economic class. If anti-social behaviour is more common among those who happen to eat less well, Collins says that food might not be the only factor.

In any case, Natural Justice wants to expand its findings by studying a larger group of prisoners. 'These findings have considerable implications and should be replicated. We may have seriously underestimated the importance of nutrition for our social behaviour.'

Does dining in the dark make food taste different?

Why not turn the lights off tonight and see? Or, for a more satisfying, professional and ultimately expensive experience, take a trip to the Dans le Noir restaurant.

The London eaterie serves its starters, main courses and desserts to customers kept entirely in the dark. Waiters are blind and the wine glasses unbreakable. The point, says Edouard de Broglie, who is behind the venture, is that removing visual cues overstimulates the other senses that contribute to taste.

'All your other senses are abruptly awoken and you taste the food like you have never tasted it before. It makes you rethink everything,' he told one newspaper. De Broglie had already opened a similar restaurant in Paris in association with a French society for the blind before coming to London.

Charles Spence, an experimental psychologist who studies sensory perception at the University of Oxford, says: 'It's partly that people will concentrate more on senses they don't usually think about. Take away the dominant sense of vision and you'll pay more attention to the sound of the food you're eating, or actually to the taste.'

This is similar to, but different from, how people who lose the use of one sense compensate by developing others. Brain scans of people who lose their sight, for instance, show large amounts of neural reorganisation, with redundant areas like the visual cortex adapted to process sound. 'If you go into a dark restaurant for a few hours you will see changes in how people perceive things, but it wouldn't be correct to attribute those to the same set of brain changes you see over weeks and months,' Spence says. 'The mechanisms aren't really the same.'

Peter Barham, a physicist at Bristol University with an interest in molecular gastronomy (the science behind cooking), says that diners could be in for an uncomfortable experience. Vision is a key part of taste, and a few drops of food colouring have been shown to bamboozle wine tasters, and fool people into eating foods they profess not to enjoy.

'I can't really see that level of confusion adding to the enjoyment of a meal,' he says. 'I would find it unlikely that you would gain much from this. You would certainly get a very different eating experience. But different is not the same as better.'

Why didn't George Best's Antabuse implants stop him drinking?

They can't; only George Best could have stopped George Best (who died in 2005) from drinking. The implants act only as a deterrent. 'They are purely a chemical terror-based medication,' says Dr Kris Zakrzewski, a private consultant who has fitted the implants in hundreds of patients.

'They work on the assumption that people know if they drink they would have some pretty violent physical effects.' But the effects differ from patient to patient and Zakrzewski says up to half of all patients experiment with alcohol, to see what they can get away with.

The implants are fitted under the skin of the abdomen, directly above the stomach. They release a steady amount of a chemical called disulfiram, which interferes with the way alcohol breaks down in the body. Alcohol usually oxidises all the way to carbon dioxide and water, but disulfiram blocks this reaction, causing a poisonous intermediate called acetaldehyde to build up. With even small amounts of alcohol this can cause headaches and vomiting; more severe reactions can lead to heart failure, coma and even death. 'They will get drunk, but they will also get very sick and sometimes even collapse,' Zakrzewski says.

Sports & Games

Is amateur boxing safe?

You're in the ring with an opponent whose main objective is to punch you, and punch you hard. So amateur boxing is not inherently safe.

But as Keith Walters, chairman of the Amateur Boxing Association, says, what sport is completely safe? 'Boxing is a dangerous sport. So is mountaineering, and so is cricket.'

While a string of tragedies and injuries has tainted professional boxing, steps have been taken to make amateur boxing safer. First, amateurs fight fewer rounds, and each round is shorter. When a professional steps into the ring, he or she must be prepared to fight for 12 three-minute rounds. Amateurs fight at most four rounds, each of which lasts just two minutes. 'The bottom line is the fights are shorter, so they get hit less,' says Walters. 'Nobody's ever been seriously injured in a two-minute round contest.'

Boxing gear can also be used to reduce injuries. Professionals use 6 oz gloves in categories up to lightweight, and

8 oz gloves above that weight. But all amateur boxers use 10 oz gloves, which are more tiring to hold up and sap the power from punches. 'You just can't punch as hard with the bigger gloves,' says Walters.

Headguards, which must be used by amateurs, also help to reduce injuries, but aren't worn to prevent the kind of devastating brain damage suffered by Michael Watson in his 1991 WBO world championship bout with Chris Eubank. 'They are to stop cut eyes when boxers clash heads, but that's all they do,' says Walters.

According to the ABA, the way the sport is scored also encourages less dangerous fighting. A boxer will commonly win on points, and to score a point, three of the five judges must push a button declaring a landed punch within one second of each other. The button-pushing system makes it pointless for a boxer to throw a flurry of quick punches because the judges simply won't be able to keep up. 'They can't hit the buttons fast enough – and when boxers get into the ring they know that,' says Walters.

How gruelling is Formula One?

It's hot, it requires unbroken concentration, and every time you accelerate, hit the brakes or corner, your body is subjected to draining g forces.

Among the physical demands on the drivers, g force, the multiples of gravitational force the driver experiences, is one of the most tiring.

In corners, F1 drivers typically experience sideways forces of

4g, and about 5g on braking. As they accelerate out of the corners, they'll feel a 1.5–2g pull. 'It's extremely exhausting,' says John Nixon of the motorsport group at Cranfield University.

Since Ayrton Senna was killed in a crash in 1994, changes in regulations have made tracks slower and made it tougher for cars to corner at speed. But one way to make tracks slower is to add more corners, so while drivers may pull fewer gs these days, the cumulative effect is still considerable.

The effect of dealing with such high g forces is noticeable in drivers. 'If you compare an early picture of David Coulthard with how he looks now, his neck and the muscles on his jawline look very different,' says Nixon. To exercise his neck muscles, former F1 driver Damon Hill fashioned a Heath Robinson-style system of ropes, pulleys and weights to pull sideways on his helmet-clad head.

For most F1 drivers, g forces rarely cause physical damage, but in 1992, Don Garlits, a drag racer, was not so lucky. After covering a quarter of a mile in a few seconds, he hit the brakes and released a parachute to slow down. The g force was so intense that his retinas detached, forcing a period of bed rest while they settled back into position and ending his involvement in the sport.

Few F1 drivers will experience a worse pummeling by g forces than at Becketts at Silverstone, where a rapid series of corners pulls 4g one way, then the other, then back again. 'It's a real battering,' says Nixon.

What would the ultimate Formula One car be like?

We'll never know for sure, because Formula One's governing body, the FIA, regularly introduces restrictions to stop cars going too fast.

But engineers have their dreams. According to Frank Dernie at Williams F1, if restrictions were lifted, a car could be built to lap Silverstone 17 seconds faster than today. There's a catch, though: it might be humanly impossible for those competing to drive.

The most noticeable difference would be the number of wheels. The large rear wheels of the cars produce a lot of drag, which slows them down. A more streamlined car would have six small wheels, two at the front and four powered at the back.

Next, the car would have an underbelly shaped like an upside-down wing. This produces low-pressure air under the car which sucks it on to the road, increasing traction. A flexible skirt around the car maximises the effect. Finally, active suspension that tips the nose of the car down for grip, or raises it for speed, would help.

There is a limit to what humans can drive. The ultimate F1 car could subject drivers to forces greater than 6g in corners, making them likely to black out.

'It's sad today,' says Dernie. 'Most of the interesting technologies are banned.'

What is the best way to bat a cricket ball?

Depends whether you want to hit it out of the ground or subtly punch it through a gap in the fielding.

The latter is an issue of patience, timing and angling the face of the bat in the right way.

But if you fancy yourself more in the Andrew Flintoff school of aiming for the commentary box, then understanding physics won't hurt. The trick is to swing the bat as fast as possible and to time your swing perfectly. Sounds simple? It isn't.

'The player only has a certain amount of kinetic energy they can give to a ball,' says Martin Strangwood, who runs the sports material research group at Birmingham University. 'The bat has a certain mass and you're swinging it with a certain speed.'

A cricket ball is made of hard leather, covering a more deformable cork core. When it hits a bat, the ball deforms (the bat compresses too but not as much) and loses energy. It gains energy, however, from the swing of the bat, and the trick for a batsman is to transfer as much energy as possible into the ball to make it fly further away. A fast bat may transfer a smaller portion of its energy to the ball, but it would tend to have more energy in the first place.

In addition, even batters like Flintoff need perfect timing. During the first part of a swing, the bat will accelerate. It reaches a maximum speed in the middle of the swing and then begins to decelerate on the outswing.

'What you want to do is get your maximum speed where

you hit the ball,' says Strangwood. 'If you are through the stroke too quickly, that means you're slowing down when you hit the ball and you're also hitting the ball up, rather than straight.'

The other option is to use a hollow bat. These would deform more than traditional solid bats, resulting in a better energy transfer – Strangwood says up to 15% more – between bat and ball. Howzat?

How do I become a synchronised diver?

It helps to be Chinese. Not because the Chinese have a physical predisposition to excel at the Olympic discipline, but because the country has such a big diving team. This means that when it comes to synchronisation, the Chinese have a greater chance of finding an ideal partner.

For while British silver medalists in the 10 m platform event Leon Taylor and Peter Waterfield may be skilled, graceful and very hard working, they are far from ideal partners – as becomes immediately obvious when they compete together. Taylor towers over his colleague.

'We have the same problem with all our synchronised divers,' says Paul Hurrion, a biomechanics expert with British Swimming. 'If they've got the same build, the same height and have the same strength and power, then it makes life a lot easier. But it boils down to them being the best in Britain by a long way; we're not like the Chinese with lots coming through who they can pair somebody with.'

For ill-matched duos to succeed, they need to use a little

physics. 'Once a person leaves the board he becomes a projectile and is subject to the forces he has delivered to himself prior to take-off,' says Eric Wallace, a sports scientist at the University of Ulster.

Gravity will ensure that both divers enter the water at the same time, provided they modify their simultaneous take-offs to reach the same height. The problem is their rates of turn: taller divers naturally spin slower, so need to reduce their moment of inertia. One way to do this is to alter their take-off. Or they could tuck up tighter than their colleague and open their legs slightly during the manoeuvre – though too much of this makes the two dives look different.

'You try to marry it up as best you can, but there are certain bits you can get away with in the middle because it just happens so quickly,' Hurrion says.

What happens to your body after a marathon?

Your immune system has taken a battering, your muscles are torn in lots of places and you won't fully recover for a few weeks.

The severity of the damage and speed of recovery after a long-distance run depend on how fit the person is to start with. For the casual runner who has spent several months training properly, the physiological damage during the marathon would have started as a gradual congestion of waste products in the body. 'Your tissues are asked to do something that they're not normally asked to do at that speed at that number of repetitions at that distance,' says Neil Black, head of physiotherapy at the English Institute of Sport.

The fatigue leads to soreness and tightness in the muscles. This means that the runner starts to move slowly and inefficiently. 'When they slow down, they will change their gait slightly,' says Clyde Williams, a physiologist at Loughborough University. 'They're changing to a recruitment of muscles that have not been used for training and that's when you get further aches and pains.' In addition, about halfway through the race, the constant pounding of the feet on the roads starts to cause pain in the joints.

At about 20 miles, levels of glucose in the bloodstream start to drop and the stores of carbohydrate energy in the body are almost depleted. Runners will become more aware of the distress signals that the various parts of the body are sending to the brain. 'The distractions of the crowds and the bands and the cheering become less of a distraction and the focus goes more on the body,' says Williams.

Dehydration is also a risk. Runners drink water en route to replace the fluid lost through sweat but can't fully replace it, simply because of the time it takes to get water through the digestive system into the bloodstream.

All this adds up, effectively, to major trauma. After the race, runners are left with microscopic tears in their leg muscles, which leak proteins such as myoglobin into the bloodstream. 'The body's defence mechanism will see these tears as damaged tissue and will set up an inflammatory response,' says Williams. Part of this response is that free radicals are released, which also attack the tissue. This leads to the familiar soreness.

Runners often pick up a cold or an infection afterwards, as

the immune system tends to be suppressed for several hours after the marathon.

It takes weeks for the body to return to normal. 'A lot of people, after a week, feel reasonably well recovered but it would be very unlikely they truly had recovered at a physiological level,' says Black.

Does grunting help tennis players with their game?

From a psychological point of view, perhaps it does. But as the grunts get louder (Wimbledon women's champion Maria Sharapova reportedly screams at about 100 decibels per shot, a volume similar to a small aircraft taking off nearby), some have called for players to be banned from doing it, arguing that the noises are unbecoming of such a genteel sport.

For some players, grunting is an inherent part of their game. 'The timing of when they actually grunt helps them with the rhythm of how they're hitting and how they're pacing things,' says Louise Deeley, a sports psychologist at Roehampton University. 'It may be that their perception is that if they grunt, they are hitting it harder. It's going to give you confidence and a sense of being in control of your game.'

And world-class tennis players such as Serena Williams are not alone in their loud grunts. 'When people are exercising on their own, they'll use things like how they breathe in and out as a rhythm,' says Deeley. 'They may make similar kinds of noises to [the grunts].'

But the exact reason for the noise remains unclear. There might be a physiological advantage. 'If you're looking at reflexes in the legs and you ask someone to clench their jaw, then believe it or not, the reflexes in their legs get brisker,' says Bruce Lynne, a physiologist at University College London. 'That's a well-known problem called re-enforcement.'

But tennis is a fast-moving game, so re-enforcement is unlikely to be at the root of the grunts.

Perhaps grunting is a tool to distract opponents? In some martial arts, breathing exercises are used to train fighters to make a short, sharp grunt – known as the kiai – to threaten or intimidate opponents as well as helping to focus the fighter's own movements.

The problem with this idea, however, is that professional tennis players should be well schooled to ignore distractions. 'If they're taught how to screen out things that are distracting, their level of attention will be at the right place on the right things,' says Deeley.

Banning the grunts might cause a stir for the players. 'They may feel, on the surface, that this is going to be a distraction to their game, that it is part and parcel of what they do,' says Deeley.

But she adds that it would be easy enough for them to unlearn their habits. 'You need to get over that psychological barrier of thinking that's going to be an effect.'

How slow can you waterski?

Not slow enough to tear up and down Lake Windermere, that's for sure. The speed limit on the lake is now just 10 knots (11.5 mph) and that means, lakeland enthusiasts hope, that there will be peace, silence and the occasional slap of wave against thwart or transom. But the waterskiers of Windermere have been up in arms (and down in the mouth). There is no way to waterski at speeds lower than 10 knots.

'The reason you can waterski is that your weight is being supported by your forward movement across the water, just like an aircraft wing supports an aeroplane,' says Martin Renilson of the defence research agency Qinetiq. 'If you slow an aeroplane down slow enough, eventually it falls out of the sky. The same thing applies to a waterskier: if you are slow enough, you just cannot stay up, you don't get the lift.' A skier would get that sinking feeling at roughly 15 mph. You could redesign waterskis, he says: make them more buoyant and a lot bigger. But then it would be a very different sport. Sailing dinghies on Lake Windermere are unlikely to break the speed barrier. But windsurfers have been clocked at more than 40 mph.

What's the lowest height you can parachute from?

A near-suicidal 100 ft (30 m) seems to be about the limit.

While soldiers of the parachute regiment balk at exiting a plane below 250 ft, those involved in an extreme form of parachuting called 'base jumping' regularly leap from further down. In 1999 a base jumper (base stands for building, antennae, span and earth) parachuted from the 100 ft statue of Christ that overlooks Rio de Janeiro. And in 1992 one leapt a similar height from the Whispering Gallery inside London's St Paul's Cathedral.

Base jumpers need their parachute to open very quickly, and so do without devices used by skydivers, who travel faster and want their parachutes to emerge more gradually. The canopy is the same, but jumpers fold theirs for quick release and don't use a 'slider' around the lines to control opening. They also often use an oversized pilot chute, already out before jumping, meaning it immediately grabs the air to pull open the main canopy in time to guarantee a safe landing.

How big a fall can a person survive?

While even short drops can be lethal, people have survived horrendous falls. In 1972, Vesna Vulovic, a cabin attendant, survived a 10,160 m fall when the DC-9 she was in exploded over what is now the Czech Republic. A 102-year-old woman survived after toppling from her fourth-floor balcony in Turin. Fortunately, her fall was broken by a children's playhouse.

In very high falls, bodies can reach terminal velocity, the speed at which air resistance becomes so high it cancels out the acceleration due to gravity. Once at terminal velocity, you can fall as far as you like and you won't gather any more speed.

Vulovic undoubtedly reached terminal velocity before hitting the ground, but it is hard to achieve when falling from a building. 'A free-falling 120 lb [54 kg] woman would have a terminal velocity of about 38 m per second,' says Howie Weiss, a maths professor at Penn State University. 'And she would achieve 95% of this speed in about seven seconds.' That equates to a fall of around 167 m, which is nearer 55 storeys high.

Falls can kill by inflicting damage to any number of vital

organs, but the most common reason is due to a key artery's route through the body. 'Most people who fall from a height die because they fracture their spine near the top and so transect the aorta which carries blood out of the heart,' says Sean Hughes, professor of surgery at Imperial College London.

Landing on your side might be the best way to survive a fall, adds Hughes. It doesn't take much of a fall to cause damage. 'From a height of 3 m you could fracture your spine,' he says. 'At around 10 m, you're looking at very serious injuries.'

How can I learn to hold my breath like a freediver?

It's not as difficult as you might think. Experts say that even non-swimmers can quickly learn to hold their breath for the three minutes and 38 seconds it took Tanya Streeter to break the world freediving record.

'You hold your breath and the body gets used to the sensation,' says Simon Donoghue, a physiologist at Oxford University.

Try holding your breath and your body will gasp for air because of three things: a shortage of oxygen, a build-up of carbon dioxide and the complaints of stretch receptors around the lungs. These sense each intake of breath and tell the brain when one is overdue. The secret to not breathing is to blunt the impact of these receptors, and one way to do this is to practise taking breaths through tightly pursed lips. This stretches the lungs for longer than usual because of the time it takes to fill them.

Donoghue learned this technique from Streeter when she visited his laboratory for tests. 'After she'd been here [we] were practising it and within a few days we could hold our breath for three to three and a half minutes,' he says.

The lab tests, in which Streeter held her breath for so long that she turned blue, convinced the researchers, Donoghue says, that she has some genetic advantages that allow her to stay submerged for so long. 'The [blood] oxygen levels she gets down to are something I've only seen in people who have had a cardiac arrest,' he says. While most people have a blood oxygen level of about 98% and anything below 80% is considered dangerous, after five-and-a-half minutes without breathing Streeter's went much lower. The machine's measuring range goes down to 50%, which is itself 'not really compatible with human life,' Donoghue says. 'Tanya went down off the scale.'

How dangerous is backstroke?

According to the Royal Society for the Prevention of Accidents, some 19,000 people were hospitalised in 2002 by accidents at British swimming baths, 1,800 of whom hurt themselves colliding with somebody else. Sadly the statistics don't indicate how many of these unfortunate bathers were in the water at the time, or what stroke they were doing.

Which means the decision by a swimming pool in Blackburn, Lancashire, to ban backstroke from its crowded lanes requires a little lateral thinking to understand. Perhaps the authorities fear a visit from Thomas Rupprath, the German

50 m backstroke world record holder, who is capable of careering head-first into the wall at more than 4.5 mph – well above the average walking speed.

Or perhaps they know their physics, and have calculated that the angular velocity of a backstroke swimmer's swinging straight arm can combine with the moment of inertia of a well-toned rotating shoulder to cause more than a splash.

'Backstroke is going to have one of the fastest velocities for the hand coming down,' says Matthew Pain, a sports biomechanics researcher at Loughborough University. And should the hand come down on the side of the pool, then the impact can break fingers.

The leisure officials in Blackburn seem more concerned with the chances of other swimmers being struck by a forceful stray hand. Ironically perhaps, fellow backstrokers – swimming face up – face the biggest danger. 'Being hit on the back of the head wouldn't hurt much but I'm sure if you poked them in the eye it wouldn't be very nice,' says Pain.

Why do tennis players check the ball before serving?

It all comes down to hair. The hair on a new tennis ball tends to be smoothed flat, while a ball that's been knocked around a bit will be more fluffy. Tennis players may check three balls or more before serving so that they can select one smooth ball and one fluffy ball. The smooth ball is used for the first serve. Because the hairs are flattened down, the ball travels faster than an older ball, which should make it harder to return. But the gain in speed comes at a cost. 'The benefit is counteracted by less accuracy because you get less grip on the ball when you hit it,' says Jan Magnus, of Tilburg University in the Netherlands. Should the first serve go astray, the player will use the fluffier ball for their second serve. Although these move slower, they are easier to control and so the player is less likely to concede a double fault.

Magnus and his colleague Franc Klaassen, of Amsterdam University, have analysed 100,000 points played at Wimbledon between 1992 and 1995. Their latest study looked at how effective serves are. They found that even top professionals often have a bad serving strategy.

'You can't make your first service too easy, because even though it'll go in every time, it'll be returned too easily. But equally, you see people using an enormous first service that almost never goes in. You have to find the optimum in the middle.'

The judgment is complicated by having two serves. Magnus found that if a player lost form and started missing a lot of first serves, they often over-compensated by making the second serve too easy to return.

'This is the most common error,' says Magnus. 'They are too afraid of double faults, but double faults are not a bad thing. There's a big misunderstanding about that.' Players who never concede double faults are not pushing themselves enough. 'If you play to your limit, you will occasionally go over the line and get a double fault. But if you never go over it, you're too far away from it,' he says.

Animals &
Plants

How does an elephant give birth?

It gets its head down for a long gestation, 21 months, after which it pushes out a baby weighing a whopping 336 pounds. Whipsnade Zoo's first ever Asian baby elephant was born on 16 March 2004 and, as usual, the baby arrived between the mother's legs, head first. 'So its thud, if you want, is only two to three feet,' says Wayne Boardman, head of veterinary services.

'It's covered in amniotic fluid and that hits the ground with a bit of fluid cushioning, in a way, and then the water bursts everywhere. Within a couple of hours it should be standing up and starting to suckle the mother. She will smell it and blow on it. Lots of them will kick it. It's a gentle control kick to move it out of the sac, so it won't be suffocated by this quite thick membrane.'

The baby was a heavy responsibility. Its tiny trunk might have got in the way at the moment of birth. The pregnancy was followed by ultrasound scan. 'At about four months you can pick out the trunk on ultrasound,' Boardman says.

The baby, named Aneena, started off about as big as a dot. 'The same size as any mammal really. And then it was 24 stone or 149 kilos by the time it popped out.'

Do cats and dogs need sunscreen?

'Some do,' says Caroline Reay, head vet at the Blue Cross in the London borough of Merton. Figures show that sales of pet accessories such as nail varnish, highlights and sunscreen are booming. But is this just vanity?

Nail varnish and hair highlights are never going to be necessary, but sunscreen can be quite important. 'Sunburn is more of a problem for cats,' says Reay. If a cat is burnt often it can eventually get skin cancer. Dogs are less likely to sit and soak up the sun, but some are vulnerable. 'White dogs need to have their hairless areas protected and breeds with short hair, like bull terriers, are susceptible,' she says.

But why not just use your own sunscreen? 'Cats and dogs will lick their coats so the sunscreen has to be safe to ingest,' says Reay. Human sunscreen can contain nasties such as zinc, salicylate and PABAs, all of which are pretty unpleasant to eat.

How do you decide whether an animal is dangerous?

By poking it with a stick or, more accurately, finding out how it would react if you did.

Biologists, zoo-keepers and other experts have got together to work out the rules on which wild animals can be owned without a licence.

The panel pondered each animal's armoury, its inclination to use it, the harm the creature could inflict on a child and how it might behave if it escaped.

The experts agreed that scorpions and several types of poisonous snake should be added to the controlled list, but considered emus, porcupines, bengal cats, raccoons, and sloths harmless enough to be taken off. So, too, the Brazilian wolf spider – not to be confused with the Brazilian wandering spider, one of the most venomous creatures to scurry the Earth, for which you still need a licence.

Other creatures were borderline, including the red (lesser) panda. 'It's not much bigger than a raccoon so logically we shouldn't have listed it,' says Jim Collins, a zoologist on the panel. 'But we kept it on because it's so sweet-looking that if one was found by a child, the child might try to approach it and get bitten.'

The expert panel was unable to agree on dwarf crocodiles and their Latin American cousins, dwarf caiman, which can grow up to 6 ft long, about a foot of which comes with razor-sharp teeth. 'They couldn't inflict any more injury than the average-sized dog but the government would have a hard sell to say, "OK, your next-door neighbour can keep a crocodile without having a checkup",' says Collins.

Edwin Blake, who works with dwarf crocodiles at Edinburgh Zoo, doesn't recommend them as pets: 'When dwarf crocodiles attack you they really mean business,' he says.

'Rottweilers can be quite calm, but dwarf crocodiles have one aim and that is to cause injury.'

Why do sea mammals beach themselves?

It's a mystery. But there are lots of possible reasons, says Philip Hammond, of the Sea Mammals Research Unit at St Andrews University.

Whales are stranded on beaches around the world on a depressingly regular basis. And pilot whales, which were involved in an incident on the coasts of Australia and New Zealand, in which 168 animals including dolphins died, have a higher than average propensity to strand themselves.

It could be disease. 'Cetacean brains harbour lots of parasites,' says Hammond. If these parasites overwhelm the animal, it could get confused and swim into shore.

Or it could be a mistake of navigation. A lot of sea mammals use the Earth's magnetic field to find their way around. Where the field's contours are at right angles to the shore, the animals following it could find themselves hitting land unexpectedly.

These reasons might explain individual strandings of whales or dolphins but it doesn't explain the Australian incident, where a large number of animals were involved. Hammond says that some species of sea mammal swim in large groups with a leader to guide them. If the leader makes a mistake, it could mean the end for them all. Another theory is that low tides around the south of Australia, combined with an increased flow of food near the shores, bring the cetaceans closer to land.

Finally, there's the possibility of human interference. There have been accusations that seismic tests to find oil and gas in the ocean near Australia may have distressed the animals, causing them to beach themselves. In the past, military sonar tests in the Atlantic Ocean – which are extremely loud to the sea mammals, who also use sound waves to navigate – have been blamed for strandings in the Canary islands and in the Bahamas. Hammond says that human interference can also explain why animals of different species are sometimes seen stranding at the same time.

How do you make a shatter-proof conker?

You could soak them in water to soften them, so that when they break they safely split. Or you could choose a young, soft conker. Just don't expect to win many matches.

Normal conkers don't often produce nasty shards, according to Hugh Dickinson, a botanist at Oxford University.

'The fresh conker is quite highly hydrated. It's got lots of fat inside it, and it more or less cleaves and splits and the bits that fly off aren't like shrapnel at all,' he says.

If you've got a conker that you haven't put in the oven for a long time and you haven't embedded in Araldite, it doesn't shatter.'

The seed coat of a conker is made from several layers of cells full of heavy waxes and lignins and is designed to protect the more delicate innards when it falls from a horse chestnut tree. If left to dry, it will eventually become harder (and therefore marginally more useful in a game) but it will also become more brittle. That might bring the possibility of flying shards.

Dickinson says that, in principle, it might be possible to genetically engineer horse chestnut trees to produce some sort of super-conker that could win matches with no risk.

'It's an incredibly big business as far as crops are concerned to balance the things in the seed,' Dickinson says. Scientists often add more or different types of fat or proteins to seeds.

Adjusting conkers for optimal physical characteristics would be a challenge few scientists would want to take up.

For a start, it would take an enormous amount of time: several decades for a horse chestnut tree to grow from a conker and start producing its own seeds. 'You do an experiment and leave it for the next generation, that's the trouble,' says Dickinson.

What's a pet psychologist?

Actually they prefer the term companion animal behaviour counsellor. 'Aggression is the most common complaint we get from clients,' says David Appleby, an animal psychologist who runs the Pet Behaviour Centre in Defford, Worcestershire, and helped to found the British Association of Pet Behaviour Counsellors.

The clients, of course, are the owners and not the animals themselves, and it's the humans that end up being given most of the homework to try to improve the behaviour of their pets. Sometimes they are asked to stand in a dog's basket to show the animal who's boss. Others may act as stooges to create confrontational situations and retrain the animal not to bite. Drugs can also be used, and anti-depressants and pheromone sprays are often prescribed for dogs and cats who have 'separation problems' when their owners desert them to go to work.

What should you do if you meet a tiger in the street?

Run. Just like the residents of New York did when Apollo, a seven-year-old, 450 kg white Bengal tiger escaped from a circus and took a stroll through Queens.

Screaming picnickers fled with their children in tow and cars crashed into each other on a busy road nearby. But Apollo remained calm in the melee, eventually being coaxed back into his cage by his trainer.

If Apollo were any old tiger, he might have reacted differently. 'It would be extremely frightened, therefore its behaviour would be very unpredictable,' says Miranda Stevenson, director of the Federation of Zoos. If Apollo wasn't used to people, he would probably have gone straight into hiding. But as he was a circus professional, he was probably happy to just stroll around the parks.

However, this doesn't mean that Apollo is a safe animal. 'The fact that it was used to people could make it more dangerous,' says Stevenson. 'It's like the tigers that turn into maneaters in Asia; the less fear they have of people, the more likely they are to attack them.'

Either way, if you were to meet a big cat like Apollo in the street, bear in mind that this is not a stray pet. There's no point staring the animal down or trying to convince it that you are the boss, for example. And don't corner the tiger. 'If it's afraid or it's in an unpredictable situation it could just attack because it's been cornered,' says Stevenson.

If it's you who is being cornered, your only option is distraction. Stevenson says that if you had a lump of meat, throwing it in the opposite direction to your planned escape route would be a good idea.

How many swallows make a summer?

Proverbially, one is not enough. 'I am wondering if the answer should be two. Or should it be 17?' says Tim Sparks of the Natural Environment Research Council's station at Monks Wood in Cambridgeshire, and part of the UK Phenology Network. Phenology is the science of when things happen. One swallow does not make a summer, because individual birds may overshoot, arrive too early, have been blown north by accident, and they tend to retreat swiftly if conditions are harsh. The presence of two swallows might well be a sign of preliminary nesting behaviour. The presence of three, or 17, suggests that they mean business. On the principle that nature usually knows best, phenologists have kept records for six years, including the first frogspawn, bud bursts, snowdrops and the earliest cuttings of the lawn.

Weather records have been kept in central England for nearly 350 years. According to the Met Office, six of the seven warmest years since 1659 have occurred since 1990. A pattern of milder winters and longer summers is now difficult to mistake.

The phenology network had reports of the first snowdrops and frogspawn before Christmas 2003. During December, 72 people cut their lawns – one of them on Boxing Day. The consensus, Sparks says, is that spring is arriving on average three weeks earlier than 30 years ago, and autumn at least a week later than it did.

Traditionally, according to the Royal Society for the Protection of Birds, the first passeriformes should be sand martins in March, followed by house martins, with the swallows appearing in the second half of April. But the presence of early swallows and the first house martins seems to confirm the bigger picture of a changing Britain.

Why do tortoises live so long?

Heard of 'live fast, die young'? Well, the opposite is true as well. The secret of tortoise longevity is more low metabolism than low speed, but the two are linked. As a general rule, animals with a high metabolic rate die early, and those that burn energy more slowly plod on for decades. The more active the animal, the higher its metabolic rate, as it has to burn energy to maintain its activity.

Take the shrew: its life is a blur, and few live to be two years old. Likewise hummingbirds. Giant tortoises on the other hand, which burn energy at a far lower rate, can crawl into an eighteenth decade. Metabolic rates differ dramatically between species. According to Jared Diamond, the UCLA-based physiologist, metabolic rates vary by 10 million times among vertebrates alone.

Scientists measure animals' metabolic rates by making them walk on treadmills and measuring the oxygen they gulp. 'It's trickier with some animals than others,' says Armand Leroi, an evolutionary developmental biologist at Imperial College London.

While scientists know longevity increases as metabolic rate drops, there is still some controversy as to why. Many believe that ageing is linked to the production of free radicals, reactive particles that are released into the body as it burns fuel. 'There's plenty of evidence to suggest that free radicals damage proteins and DNA,' says Leroi.

But metabolism is not the be-all and end-all, says Leroi: 'If you were to slow down your metabolism, you'd still die from any number of things.'

Are chow chows the most stupid dogs in the world?

Not quite, but they are certainly up there. Newspapers reported that a chow chow from Bispham, Lancashire, was badly injured when it leaped from a second-floor window after being startled by a plane flying over its house. But this doesn't begin to reveal the true mental deficiency of the breed.

Judging the intelligence of any animal is difficult but according to Stanley Coren, a psychologist at the University of British Columbia and author of *The Intelligence of Dogs*, it can be done.

Coren wrote to all the registered dog judges in North America and asked them to rank 110 breeds of dog by their 'working intelligence', a measure of how well dogs learn. Because some breeds were evenly matched, Coren ended up with 79 ranks of canine intelligence. 'The chow chow turns out to come in 76th out of 79 ranks,' says Coren. 'What that means is that there is probably furniture out there that is more trainable than chows.'

Coren's study showed that to train a chow chow to do something, like sit when told to, it took on average 80 to 100 attempts before the dog grasped what was being asked of it. It would then have only a 25% chance of remembering what it was taught. (Only Afghan hounds, basenji and bulldogs turn out to be thicker than chows.) In contrast, the whizzkid of the dog world, the border collie, needed telling just five times to learn a new command and was likely to remember it 95% of the time afterwards.

It's not the chow chow's fault it's so stupid, though. 'It's totally understandable,' says Coren. 'Chows were originally bred as food animals. Who needs smart food?'

In China, some farms still raise chows for meat (folklore says black ones are better fried while others should be stewed). The dogs are not called chows because they make good 'chow', as is commonly supposed.

In fact, when they were first shipped to England, they arrived in boxes marked 'chow chow', pidgin English for miscellaneous merchandise. 'The customs people simply assumed that was what they were called and the name stuck,' says Coren.

What really killed off the dinosaurs?

Just as scientists thought that they had nailed down the answer, the debate has been reopened. A team of scientists has claimed that the widely accepted theory that the extinction was triggered by a huge asteroid thumping into Mexico 65 million years ago, cannot be true.

Evidence that a giant asteroid impact was the cause of the dinosaurs' demise first emerged in the 1980s. Scientists analysing ancient soils in Italy found that layers of clay from the end of the Cretaceous period, the time the dinosaurs vanished, were unusually rich in a heavy metal called iridium. Later evidence of the layer was found in other countries, including Denmark and New Zealand. The most likely cause was believed to be an extraterrestrial rock that struck Earth and showered iridium across the continents. Such an impact would have had a devastating affect on life, as hot rocks fell from the skies and dust shrouded the sun.

The theory gained credibility a decade ago when scientists declared they had found the smoking gun for the impact. A crater more than 100 km across, that seemed to date back to the end of the Cretaceous period, was discovered near a village called Chicxulub on the Yucatan peninsula.

But according to Gerta Keller, a geologist at Princeton University, the Chicxulub crater is not linked.

Keller's team analysed rock which had melted in the intense heat of the impact, been thrown into the stratosphere and scattered far and wide. They found the oldest pieces, which have the same chemical composition as molten rock in the crater, were formed some 300,000 years before the dinosaurs became extinct. Samples from the crater back up the idea that dinosaur life existed long after the impact at Chicxulub, says Keller.

'What this means is that Chicxulub is not the smoking gun that caused the extinction. What really killed the dinosaurs must have been another impact,' she says.

And so the search for the real smoking gun is on again. If, of course, an asteroid were actually responsible.

Can animals be homosexual?

Very probably, but declaring it might not be wise, as Oxford student Sam Brown discovered when he was arrested for asking a mounted police officer if he knew his horse was gay.

In his 1999 book, *Biological Exuberance*, Bruce Bagemihl documented many cases of homosexual behaviour in animals. But as researchers point out, occasional homosexual behaviour is not the same as exclusive homosexuality. 'The issue is complicated because we don't know anything about the eroticism in the heads of animals,' says Linda Wolfe, head of anthropology at East Carolina University.

Wolfe has spent much of her time studying homosexual behaviour in animals, chiefly non-human primates. The variety of same-sex interactions is as diverse as any dreamed up by humans, she says, although prudishness often prevents it being discussed. 'It's particularly bad in the US,' she says. 'There's a video some researchers made of male bachelor gorillas engaging in fellatio, but it still hasn't been shown in the US,' she says. 'As far as I know, homosexual behaviour has been observed in all manner of animals, and if pleasure is driving it, why wouldn't they?'

In 2004, Charles Roselli at the Ohio Health and Science University reported differences in the brains of gay rams. Certain groups of cells that govern sexuality in the rams' brains were different in those that preferred rams to ewes. According to Roselli, about 8% of domestic rams show a sexual preference for other males.

Bagemihl's book caused a stir when it was published, not least because of a speculative chapter about the implications of gay animals on Darwinism. According to Wolfe, the explanation may be little more than animals seeking sexual pleasure.

What should you do if a shark attacks?

Over to Douglas Herdson of the National Marine Aquarium, Plymouth. 'Really, you shouldn't get yourself into a situation where they feel threatened in the first place,' he says helpfully. 'They would normally give warning signs – if a shark starts going around and humping its back, then it doesn't like you being there and you would be silly not to slowly back away and get out of its area.'

But chances are that you won't even see it. 'On most occasions when a shark is genuinely attacking a person, the person hasn't seen the shark beforehand,' says the aquarium's Rolf Williams, who has researched sharks for years. 'They're excellent predators; you can bet you won't see it coming because they need that element of surprise.'

A great white would lurk deep underneath. If the shark saw a silhouette at the surface and decided to take it, you wouldn't see anything until it hit you. 'You swimming at the surface, particularly if you're wearing a wetsuit and flapping around, is rather like an injured seal,' says Herdson.

In any case, the advice is to stay calm. 'By thrashing away you are attracting its attention and you are making a noise like a wounded animal. Also you're not keeping an eye on it,' says Herdson.

If you're in the deep ocean, you might see sharks circling. Here, there is some more practical advice. 'Keep eye contact,' says Williams. 'If you have a stick, you hold it straight out towards the shark's eye and that tends to keep them at bay. If you haven't got anything, hold your foot out and point that at the shark's eye and they will keep one or two feet beyond that.'

If the shark is still interested, it will come and nudge you with its nose to see what you are. 'Then you would give it a good hard kick in the nose, the eye or the gills,' says Williams.

Of course, if you are stuck in the middle of the ocean and the shark is determined, it could wait until you're too tired to fight back. In that case, well, there's not much that will help.

Why are giant pandas so bad at mating?

The truth is that male pandas are shockingly rubbish at the mating game. Male giant pandas are bad at working out when a female is likely to welcome their advances, and bad at knowing what to do next if they do happen to stumble upon a willing mate. In the unlikely event that they get around to having sex, they're often too quick about the whole business to get the female pregnant.

To be fair, males have the odds stacked against them. Female pandas are only receptive to them for two or three days a year. 'Often, males don't read the signs right, and if they try at the wrong time, they get bitten,' says Susan Mainka, of the World Conservation Union's species survival programme.

The difficulty in getting pandas to mate is compounded by the natural pickiness of females. In the wild, female giant pandas usually select a mate from a group of males, but this is a luxury they are often denied in captivity: Ling Ling and Shuan Shuan (the names of giant pandas are traditionally repeated as a form of endearment) are among only around 150 giant pandas in captivity.

Estimates suggest just over 1,000 pandas remain in the wild, in three Chinese provinces. The biggest threat to their existence comes not from their difficulty reproducing, but from poachers. Although the practice has long been illegal, with poachers facing fines and up to 20 years in prison, the fact that pelts can fetch more than twice the annual income of a rural worker means it is still a big problem.

The destruction of the pandas' natural habitat, largely because of deforestation, is also taking its toll on the animals. Pandas are now confined to small, isolated populations living among bamboo thickets on steep mountain sides. Experts believe the isolation of the communities is making pandas increasingly inbred, which may be further damaging their ability to reproduce. Pandas are also in the precarious position of relying almost exclusively on bamboo (they typically munch 25 kg a day in around 12 hours) for sustenance. If bamboo dies off, as it has in the past, pandas can be left stranded with no food.

Over the years, the Chinese authorities have tried ever more desperate attempts to revive the panda population. Males in Sichuan were given Viagra to help improve their staying power beyond 30 seconds. Mainka, who timed pandas having sex in China a few years ago, says the longer they keep at it, the more likely they are to produce offspring.

Is it safe to swat a mosquito?

This question has gained unlikely prominence, thanks to Christina Coyle and her colleagues at the Albert Einstein School of Medicine in New York.

The team reported in the *New England Journal of Medicine* that a woman who died of a muscular fungal infection probably met her fate after smearing a mosquito into her skin.

Mosquitoes are carriers of fungus-like parasites called Brachiola algerae, which in rare cases can cause such infections. The verdict from the scientists is kill carefully: flick mosquitoes from your skin, don't just squash them dead.

But according to Gordon Leitch, an expert on Brachiola species at the Morehouse School of Medicine in Atlanta, Georgia, encouraging the world's inhabitants to change their preferred method of execution is a little rash.

'There have been three, at most four, infections from Brachiola ever, and I mean ever,' Leitch says. Of these, one was a simple skin infection, another an infection of the eye, unlikely to be caused by a mosquito bite. Leitch recommends the insects be dealt with swiftly. 'Swat the little buggers,' he says.

How long can a seed stay alive?

There are no definite answers here. From the research done by conservationists, the durability of a seed is known to depend critically on how it is stored: keep it in ultra-cold, dry conditions and you can expect it to stay alive for several hundred years.

Which makes the story from Israel, that scientists have grown a date palm from a 2,000-year-old seed found during archaeological excavations on Mount Masada, seem extraordinary. The Israeli team say the age of the date palm seed was verified by radio carbon dating.

But it's not the only ancient seed to have germinated: in the mid-1990s, a Chinese lotus plant grew from a seed that was dated at around 1,400 years.

At the Millennium Seed Bank – a leading centre for long-term storage of seeds, based at the Royal Botanic Gardens in Kew – initially, seeds are usually dried to between 4% and 6% moisture content. Then the seed is kept at –20°C. 'For all the wild species that live on our seed bank, we estimate that for most species we're certain of many hundreds of years [of survival],' says John Dickie of Kew's seed conservation department, which runs the seed bank.

The only way to check how long seeds really survive is to plant them in a few hundred years' time. A more practical method is to use a mathematical model, which projects what botanists know about seed survival into the future. Dickie has found that if wheat grains are kept at a constant 16°C, one grain in a thousand might germinate after 236 years. With temperatures in the high 20s, the grains would all be dead in 89 years.

The Chinese lotus plant survived so long because its seed would have been impervious to water and, by falling to the bottom of the lake in which it was found, it stayed relatively cold.

But, according to Dickie, surviving two millennia in the desert soil of the Middle East stretches the imagination. 'I would have thought the average temperature is working against you,' he says.

Do badgers spread bovine TB?

While research over the past 20 years suggests badgers are a factor in the spread of bovine TB among cattle, there is no conclusive evidence.

Steve Kestin, a scientist at Bristol University, says the links between badgers and the spread of bovine TB are strong. 'The badgers almost certainly get TB from the cattle. But it's then more than likely that the badgers give it back to the cattle. Badgers have a busy lifestyle which takes them into farmyards and all over the pastures.'

Bovine TB was endemic in the national cattle herd in the 1930s. The government began a culling policy that wiped out the disease 20 years later. Pockets of TB appeared soon after and, almost by chance, it was found in badgers that had been run over by cars.

Now, bovine TB affects more than 5,000 farms and 6% of the country's herds. The government has been looking into its spread using a method called the Krebs trial. This operates by marking out 10 km square sections of countryside and comparing three different strategies within each area: in the first patch, as many badgers as possible are culled; in the second, badgers are only killed if the cattle herd is infected (so-called 'reactive' culling); and, in the third, they are left alone. The results have been a surprise.

'In the reactive culling squares, TB instances have gone up in cattle,' says Kestin. 'As a result, they've stopped that treatment. When you cull out a population of badgers, you cause a lot of movement and re-establishing of territories. The populations outside the areas you've culled move in.'

But Kestin says the farmers' demands to kill all badgers is a 'knee-jerk reaction that has just got jerkier'. He says that there are several possible reasons for the spread of bovine TB. Other animals such as deer and hedgehogs could spread it. Also, cattle are moved around more these days. 'Farmers move infected cattle to new areas and those cattle then infect other cattle, and there's a risk that they then infect the badgers,' says Kestin.

Is it time to start culling seagulls?

No. For one thing it's impractical, because of the large numbers of birds involved. But there's also doubt about how much of a problem gulls are.

It is true that numbers are increasing and they are living further from their natural coastal homes. 'Every year, there's probably one or two cases of people having closer contact with gulls than they would like,' says Grahame Madge of the Royal Society for the Protection of Birds. 'When you consider gulls are very widespread the number of incidents is insignificant compared to the number of birds around.'

The reason for the increase in urban gull numbers is simple: food. 'Urban gulls get plenty, wild gulls don't,' says Peter Rock, a researcher at Bristol University and Europe's leading authority on seagulls.

Urban gulls appear to eat everything from rubbish in landfills to scraps on the street, and produce up to three chicks per pair a year. For wild gulls finding food at sea is increasingly difficult and they may produce one chick every 10 years.

The increase in town birds certainly causes headaches for humans. Gulls are large, make a lot of noise, produce copious droppings and can be aggressive towards people or animals.

But it seems there's no quick fix. Shooting and poisoning the birds have been mooted, but Rock says the ideas are silly without proper research into how the urban gulls live.

Madge agrees: 'The notion that you can actually solve this by culling or other measures is very short-sighted.'

Why are fractures often fatal for racehorses?

Tim Greet, former president of the British Equine Veterinary Association, says there is nothing inherently weak about horse bones that makes fractures life-threatening. 'Bones in horses heal very well – if you took 100 fractures, the majority do heal.'

But when a bone breaks, vets have to consider whether the suffering during the convalescence required is the right thing for the animal. 'When long bones break, they do so like a grenade and it may not be possible to put the bits back together again,' says Dr Greet.

Horses make for poor patients. As soon as they wake up from surgery, they will try to stagger to their feet.

'Three hours of surgery can be damaged in 0.3 seconds,' says Dr Greet. 'If you're bred to race and if your leg is broken badly so that you cannot race again, you have to look at if it's fair to put a horse through suffering during its convalescent period.'

Natural Disasters & Man-made Problems

Would Britain freeze if the Gulf stream were switched off?

The meridional overturning circulation (MOC) is the current that drives the Gulf stream, and news of it slowing down by one third in the past 12 years is alarming. Before the decline, it shunted tonnes of water through the Atlantic every second, bringing the equivalent of a million power stations' worth of heat to Britain's shores, making the conditions extremely mild for our latitude.

The scientists, led by Harry Bryden at the National Oceanography Centre in Southampton, said that if the current remained weak, it would equate to a 1°C drop in Britain's temperature. And if the current shut down entirely, temperatures could drop by 4°C to 6°C.

But the chilling effect does not take into account the fact that the world is still warming, thanks in part to greenhouse emissions. According to Richard Wood, a climate modeller at the Met Office's Hadley Centre, advanced computer simulations predict a range of outcomes into the 21st century from a negligible change in the MOC to a 50% slowdown of the current. Crucially, when those results are fed into other climate simulations, even the most severe weakening of the current fails to cool Britain, because the heat from global warming outweighs it.

'None of the outcomes gives a net cooling. All of the models show we can expect warming pretty much everywhere, apart from over some regions of the Atlantic,' he says.

Even if the current were to shut down, it would have to happen rapidly to plunge Britain into a much harsher climate. 'If the current shuts down rapidly, in 20 years or less, there could be a cooling, but if it happened gradually, such as over 200 years, the chances are global warming will more than counteract it,' Dr Wood adds.

The last time the current stopped was 8,000 years ago as the world emerged from its last ice age. Climatologists believe it was caused by a huge amount of ice that had built up on the Canadian coast melting in the Atlantic, adding a vast amount of fresh water. 'There was a lot more ice to melt at the end of the last ice age,' said Dr Wood.

What are rogue waves?

Over to Mark Stubbs, whose boat the *Pink Lady* was broken in half by a rogue wave. The disaster brought his four-man team's attempt at rowing across the Atlantic in record time to an abrupt end.

'We didn't see the one that got us because we were in the

cabin, but we did see one earlier. It was twice the size of anything else around – around 40 foot, and was a lot louder,' he says.

Rogue waves have an almost mythical status, but scientists have monitored them using satellite imaging. 'Rogue waves are abnormally high waves that are larger than you'd expect, given what you know about the energy of the sea at that region,' says Chris Swan at Imperial College London, who uses computers and wave tanks to model how rogue waves come about.

Rogue waves occur when several small waves moving in the same direction coincide and reinforce each other. But simple wave interference is not enough to produce the effect. Rogue waves arise only when longer waves push energy into shorter waves, forcing them to become much taller and steeper than average. 'As they pump energy into the shorter waves, they have to conserve energy, and the only way they can do that is to get taller,' says Swan.

The conditions that set up rogue waves mean they appear very quickly, building up over 500–800 m, before disappearing again just as quickly. But while our knowledge of rogue waves is increasing, it does nothing to help those unfortunate enough to see one. 'What can you do about them? Probably bugger all,' says Swan.

Do trees pollute the atmosphere?

Yes, just as president Ronald Reagan said in 1981. 'Trees cause more pollution than automobiles do,' he opined. A little later, environmental scientists ruefully confirmed he was partially right. In hot weather, trees release volatile organic hydrocarbons including terpenes and isoprenes – two molecules linked to photochemical smog. In very hot weather, the production of these begins to accelerate.

America's Great Smoky Mountains are supposed to take their name from the photochemical smog released by millions of hectares of hardwoods.

Isoprene serves as a catalyst, driving the rate at which sunlight breaks down oxides of nitrogen – mostly from agriculture and cars – to produce atmospheric ozone.

Ozone is a triple molecule of oxygen. High in the stratosphere it is a godsend, screening out cancer-causing ultraviolet radiation. But in the lower atmosphere it is a toxin: it causes stinging eyes, prickling nostrils and aggravates severe respiratory problems. Statisticians calculate that in August 2003 – the long hot summer that caused an estimated 20,000 deaths in western Europe – more than 500 British deaths could be attributed to ozone pollution.

But the experts say that the trees alone are not the problem. The real villain is the motor car. Trees soak up carbon dioxide, and respire oxygen, doing far more good than harm. And finally, as one forester observed: why worry about a few harmful natural chemicals? In a truly antiseptic world we would all be dead.

When will global warming reach a point of no return?

When the world warms so much that it causes irreversible damage to ecosystems. Ice sheets will be doomed to melt, species will become extinct and critical ocean currents will grind to a halt.

At least, that's the idea. Experts say that once atmospheric levels of carbon dioxide exceed 400 parts per million (ppm), a 2°C rise in global temperature is inevitable, and so is irretrievable damage. Levels of carbon dioxide now stand at 379 ppm, and with typical yearly increases of 2 ppm, the threshold could be crossed in around 10 years.

The impact of climate change is not so clear-cut, though. 'This is a mixture of political and scientific argument,' says John Schellnhuber, research director at the Tyndall Centre for Climate Change Research in Norwich. 'There's no crisp threshold. The world is a mosaic of ecosystems. Some are already severely damaged, others will be with more warming.'

Declaring a specific temperature rise as dangerous is hard to justify scientifically, adds Chris West, director of the UK Climate Impact Programme at Oxford. 'It's completely subjective. You have to ask what do they mean by dangerous and who is in danger. For some, dangerous climate change has already arrived.'

Despite the criticisms, many climate change scientists agree that a rise of more than 2°C in the world's temperature will significantly increase the risk of triggering what they call 'tipping points' – ecological changes from which the world cannot recover. But even so, such changes are unlikely to happen rapidly. A study showed that a rise of more than 3°C would cause Greenland to melt. 'What they meant was at that temperature, Greenland will eventually melt. But it would take millennia,' says West.

Ambitious agreements such as Kyoto will go a little way to slowing carbon dioxide emissions, but the prospect of stopping global warming is extremely distant. 'With current rates of economic growth, the way developing countries are requiring more energy and the rate at which we're not moving to renewables, it's hard to see how we can avoid increasing emissions.'

Is there a hole in the ozone layer over Britain?

Experts hate calling it a hole, but the protective layer is thinning fast. Ozone shields us from harmful ultraviolet rays and the more radiation that gets through, the greater the risk of skin cancer and cataracts.

Ozone depletion is a largely forgotten problem since the Montreal Protocol successfully reduced levels of CFCs spewed into the atmosphere. But the chlorine-containing compounds that do the damage take decades to degrade, and scientists say thinning of the ozone layer will probably get worse before it gets better.

Cold conditions speed ozone loss and most attention until now has been on the chillier Antarctic, where a hole in the ozone layer has opened each spring since the 1980s. Following an unusually cold Arctic winter, European scientists raised the alarm about northern ozone loss in January 2005.

Is it dangerous to store carbon dioxide underground?

It's worked for millions of years. The question is whether giant bubbles of gas pumped down there from power stations will be as stable as natural reservoirs.

A report prepared for the government warned that carbon sequestration – or capture and storage – might be the only way for Britain to reach ambitious targets to reduce carbon dioxide emissions. 'With the 60% reduction target for carbon dioxide emissions by 2050, large-scale deployment of carbon capture and storage may be needed for electricity generation from about 2020,' the report said.

The UK is well placed to develop the technology because it could exploit the oil and gas drilling infrastructure in the North Sea, where many oilfields are beginning to run dry. The oil industry has long pumped carbon dioxide from natural sources into oilfields to squeeze out remaining fuel reserves, a technique known as enhanced oil recovery.

Environmental groups and politicians want guarantees that the stored gas will not leak out over time. The world's only large-scale investigation of carbon sequestration at sea is being led by the Norwegian company Statoil, also in the North Sea. Since 1996, the company has been pumping carbon dioxide into a sandstone layer about half a mile below the seabed. The resulting bubble now contains more than 6 m tonnes.

Andy Chadwick, of the British Geological Survey, is part of the team monitoring the project. He says there is no sign of the gas escaping, though the seismic surveys used to check are only sensitive enough to spot leaks of several hundred tonnes.

More sensitive monitoring has just begun at a much smaller-scale onshore project in Texas.

Scientists from the US National Energy Technology Laboratory in Pittsburgh told the American Geophysical Union annual meeting in December 2004 how they laced the several thousand tonnes of carbon dioxide injected into a salt-water formation nearly a mile underground with fluoro-carbon tracer molecules. They will now monitor the soil above to see if the stored gas makes its way back to the surface.

Will nuclear power stations of the future be very different?

As with most technology, several decades of development means that modern nuclear power plants are a world away from the clunky behemoths in operation in Britain today.

Which just might placate some of the nuclear nay-sayers, who have concerns over the possibility of building new nuclear stations to feed the country's growing need for power. They have good reasons to worry: accidents at nuclear power stations can cause horrific damage to the environment and the issue of waste disposal has not been resolved.

But if Joan MacNaughton, the director general of energy policy at the Department of Productivity, Energy and Industry (DPEI), is to be believed, new nuclear power stations must be pressed into action if Britain is to meet its targets to cut the emission of carbon dioxide. 'We now have 12 nuclear stations providing 20% of our electricity carbon-free,' says MacNaughton in a report for incoming ministers at the DPEI. 'By 2020 this will fall to three stations and 7% as stations are retired.'

Modern nuclear stations, already in operation in South Korea and China, are a world away from the old designs. 'The main change we've seen over the last 10–15 years is the shift in emphasis from engineered to passive safety,' says Malcolm Grimston, an expert in nuclear policy at Chatham House. 'Engineered safety is where you have to have valves and pumps that have to work to deal with the potential dangerous situations.'

The alternative passive safety systems are less expensive and less maintenance is required because there are fewer moving parts.

According to Westinghouse, a subsidiary of British Nuclear Fuels Limited, which designs nuclear power stations, their most advanced designs – such as the AP1000 – are around 100 times safer than existing stations.

The emergency cooling water in an AP1000, for example, is above the reactor core. In the event of an accident, the water just falls on to the core. As it begins the cooling process, it converts to steam, hits the stainless steel barrier at the top, condenses to water and rains back down on the core.

The new nuclear power stations are also cheaper. 'The designs are much simpler, there's much less in the way of equipment,' says Grimston. He estimates that the cost of a new station is probably half as much as the batch-produced stations of the 1970s. The AP1000, for example, has 36% fewer pumps, 87% less cable, 83% less safety-related pipe and 50% fewer safety-related valves. The Westinghouse design also uses the uranium fuel 60 times more efficiently and produces just 10% of the waste of nuclear power stations today.

Are pizza ovens a major source of pollution?

Possibly. 'When people think about air pollution, they think immediately about big industrial operations, power generation, transport,' says David Santillo, an environmental chemist at the Greenpeace research lab in Exeter University. 'While wood-burning stoves are not the major source of contamination, they're an important source and one that has probably been overlooked. They are one of the least regulated sources of particulates or other chemicals.'

Burning wood releases all sorts of chemicals. Carbon dioxide and carbon monoxide are the main parts of the smoke (the relative levels depend on how efficiently the stuff has been burned). There's also the possibility of sulphur oxides and nitrogen oxides, depending on the fuel being used.

There are also tiny bits of organic and inorganic chemicals in smoke – so-called particulates. 'The smaller they get, the more dangerous they are,' says Santillo. 'The problem with them is the size of the particles – if they get into the lungs, they can cause problems, even lung cancer if you're exposed over long periods.'

Where organic matter is used industrially, for example in power stations, the smoke is heavily filtered to remove as much of the nasty stuff as possible. Domestic or commercial ovens are less likely to have such filters.

Santillo says the problems of wood-burning ovens should not be over-stressed, however. In terms of urban air pollution, transport is still king.

How do you tow an iceberg?

For decades, scientists and engineers have dreamed up schemes to bring massive supplies of fresh water to parched regions of the world.

John Isaacs, of the Scripps oceanographic institute in California, set the ball rolling in the 1950s, when he calculated that a fleet of six tugboats could haul a 30 km iceberg from the Antarctic in a few months. Twenty years later, the US scientists John Hult and Neil Ostrander said the process could be improved by carving the bergs until they resembled the bow of a ship. And then there was Prince Mohammed al Faisal, one of the sons of King Faisal of Saudi Arabia, who launched a company in 1977 to see if an iceberg could be towed the 14,000 km from Antarctica to Jeddah. To promote the idea, the prince had a two-tonne Alaskan berg hoisted by helicopter to a conference in Iowa, where some of it found its way into delegates' drinks.

More recently, Patrick Quilty, of the University of Tasmania, investigated its potential in 2001 to sate the thirst of cities including Perth and Brisbane. 'Let us dispose early of one popular concept,' his report said. 'That of towing icebergs. While it is technically feasible to develop large enough vessels and systems, towing exposed icebergs between continents is, in the southern hemisphere, out of the question.' Up to 80% of the precious water would melt into the sea en route.

German scientists may have the solution: a plastic film reinforced with a fibre web that comes in two-metre rolls and can be wrapped around icebergs to seal in their water.

Can 4×4s really cause dust storms?

They certainly don't help. Dust storms mostly affect cities on the edge of desert areas. These huge clouds of dust, which reduce visibility to less than a kilometre, measure up to 130 km wide and can wreak havoc for hours and sometimes days.

Left undisturbed, desert ground is stable thanks to a thin coating of algae, lichen or clay. But heavy vehicles break up the surface and release the powdery soil beneath. 'The problem now is that four-wheel-drive use is increasing,' says Andrew Goudie, a geographer at Oxford University. 'People are using them to collect wood or round up their herds instead of using camels. And in large parts of the south-western US, there are a lot more SUVs moving around.'

Goudie says 4×4s should be banned from areas of land susceptible to dust storms, including parts of the Sahara. 'Dust storms cause real problems. You lose a lot of the available soil, you get fine particulates in the air which can cause respiratory problems, then you get spores and fungus which create disease,' he says. 'You also get visibility problems which lead to accidents on the roads.'

Do patio heaters cause global warming?

If heating the inside of the house is expensive, then just imagine how much energy it takes to heat the outside. Some patio heaters churn out a whopping 14 kilowatts, easily the same as an entire houseful of standard electric fan heaters on full blast.

Some of the opprobrium is probably unfair – similar heaters have been used to warm large open spaces including factories and supermarkets for years – but patio versions are notably less efficient because they don't use fans. This means that lots of the heat, as every school pupil knows, rises.

The Romans had the right idea with their underfloor hypocaust system – if you want to get the most from your heat source, it needs to be below you.

'Fans would convect heat downwards, but then when you're eating I suppose you don't want the fumes,' says David Reay, a consultant engineer and heat transfer specialist in Whitley Bay.

'It would have to be purified somehow or it would be like sitting in the exhaust from a gas fire.'

No fans means that the heaters must rely on radiation to shift thermal energy – and for this to happen effectively they need to get very, very hot themselves. 'The wind will convect on to the surface and cool the surface down, so it will inhibit the radiative heating,' Reay says.

How many people can the Earth support?

It's an old question. Two hundred years ago, Thomas Malthus said population would race ahead of food supply, but he wasn't the first. The early Christian writer Tertullian said (around AD 200, in *De Anima*): 'We are burdensome to the world, the resources are scarcely adequate for us . . . Truly, pestilence and hunger and war and flood must be considered as a remedy for nations, like a pruning of the human race becoming excessive in numbers.'

That was when the population of the whole planet was maybe 100 million or so. We reached the first billion mark by about 1850. By 1950, it was about 2.5 billion. In less than one short lifetime, this figure doubled. It passed six billion in the late 1990s. Note that humans took 150,000 years to get to the first billion. The most recent billion arrived in just 12 years.

Nobody knows how many people the planet could hold. The UN predicted that fertility would decline and longevity would increase until the global population stabilised at nine billion in 2300. Some optimists have argued that the planet could support 1,000 billion; others look at what is happening right now and wish that it had stayed at ancient Roman levels.

Joel Cohen, the Rockefeller University population biologist, argues in a 1995 book (*How Many People can the Earth Support?*) that it isn't a question like 'How old are you?', which only has one answer at any one time. Cohen argues that you could fit one billion people each a metre apart, into a field 32 km square. So everybody in the world would fit easily into Yorkshire. But it takes 900 tonnes of water to grow a tonne of wheat, and there is only so much water, so much land and so much sunshine. Human action has its own 'ecological footprint': there has to be so much land to provide food, clothing, shelter, medicines, building material, fresh air and clean water for any one human. It takes, according to some calculations, 2.1 hectares of land and water to provide for one average human. The important word is: average. The American footprint is about 10 hectares. So if all humans lived at US standards, we'd need another four Earths.

Where does all the water go?

More than 3 bn litres of clean water flood out of leaky or burst pipes across Britain every day, according to figures from the industry watchdog, Ofwat. But where does it all go?

'Most leaks occur in cities and the majority of that water finds its way into sewers,' says Ian Barker, head of water resources at the Environment Agency. 'Walk through any of those large Victorian brick sewers and water is always dripping in through the roof. A lot of that is down to leaks.'

Because cities are usually built along estuaries or near the sea, water that seeps into sewers is often discharged into those estuaries or along the coast, where it is effectively lost: to keep purification costs down, water companies tend to draw their water from pristine watercourses inland.

Many years ago, water lost through leaks and burst pipes beneath cities – which can be as much as 20% of that supplied – was easily recovered because it simply percolated down into subterranean aquifers. But industrialisation has left a legacy of contamination and few companies now draw water directly from urban aquifers.

Water companies agree an acceptable 'economic limit' on the amount of water they can lose to leaks and bursts with Ofwat, based on the assumption that chasing every leak costs more than letting some water escape. The acceptable losses vary, depending on the age of the pipes – up to 100 years in London – and the pressure the water is pumped at. But typically, water companies lose around 150 litres of water each day to every property they supply. 'That's the equivalent usage of one extra person per property,' says Barker.

Leaks aren't all bad news, though. Without them, experts say, urban trees would be more likely to wilt and aquifers beneath cities would remain polluted for hundreds of years.

How can you predict volcanic eruptions?

Before a volcano erupts, there will always be warning signs.

Bernard Chouet, a scientist with the US Geological Survey, described how he was able to predict volcanic eruptions by detecting a particular type of tremor in the ground. These 'long period events' were sure signs, he said, that pressure was building up inside the volcano.

Bill McGuire, director of the Benfield Hazard Research Centre at University College London, says this type of seismic monitoring is standard. 'There are lots of different methods of monitoring volcanos now but the two ways that unequivocally tell us that a volcano's getting ready for eruption are still the old ones – earthquake activity and ground deformation,' he says.

Before a volcano erupts, magma (molten rock) rises towards the surface, breaking rock en route. As more pushes up, the rocks around it vibrate. This results in earthquakes that can sometimes cause damage to buildings but are nothing like the big tectonic quakes seen on the west coast of the US or in Turkey.

Any sudden change in quake activity around an active volcano will, hopefully, give scientists enough time to sound warnings.

The magma also causes ground deformation. 'As magma rises into a volcano, it has to make space for itself and that means that the ground has to swell,' says McGuire. Every year, his team travels to Mount Etna to measure the relative positions of several reference points using sophisticated GPS recording receivers. Any changes in position – of the order of tens of centimetres – could mean the onset of an eruption.

David Rothery, a researcher in the volcano dynamics group at the Open University, uses another method of detecting potential eruptions. Many volcanos have craters at the summit but it is usually too dangerous to put instruments there. He peers at craters from space using satellites that measure infrared radiation, and looks for any sudden changes in heat activity. He says that, depending on the volcano, scientists will get anything from several months' to a few days' notice of an eruption.

In theory, then, eruptions shouldn't cause casualties. But this always depends on the evacuation plans for danger areas. Before Vesuvius erupts again, for example, 600,000 people will have to be evacuated, possibly at only a couple of weeks' notice. 'It's a matter of political will, it doesn't come down to the science in the end,' says McGuire.

How do you know when a tsunami is coming?

Many survivors of the Indian Ocean tsunami talked of the sea retreating dramatically, leaving fish flipping around in the sand, just before the first huge wave came smashing in. Although it does not precede every tsunami, such a retreat is common – a result of sea water plunging into the huge pit the earthquake creates in the ocean floor.

Sadly, the eerie spectacle of the vanishing sea is often a draw to those who witness it, instead of the alarm bell it should be. The next sign that a tsunami is coming is the wave rising high above normal sea level as it reaches shallower waters.

The Indian ocean tsunami was caused by such a huge earthquake that many would have felt the tremors before the energy of the quake was pushed into the oceans creating the tsunami. 'Although the epicentre was just off the coast of Sumatra, the actual rupture stretched up as far as the Nicobar islands,' says David Tappin, a tsunami expert at the British Geological Survey.

Even seismologists have trouble determining exactly when a tsunami is about to be unleashed, though. While sensitive seismometers will pick up the vibrations caused by earthquakes, not every huge earthquake causes a tsunami.

Early warning systems are effective only when people get enough notice and know what to do. The tsunami hit Sumatra too fast for any warning to be much use. Low-lying islands had little chance, lacking the high towers that countries such as Japan have built to provide temporary sanctuary. 'You need different plans that are specific to every region,' says Tappin.

What causes twisters?

You need a very cold, dry air stream to crash into some warm, humid air. The boundary between the two 'fronts', as meteorologists call them, is where the action happens.

Buildings can literally explode if a typical 200 mph twister flies past as the pressure inside this spinning cloud is much lower than outside it.

But conditions have to be just right for the storms to happen and, unfortunately for the people living there, the American Midwest – 'tornado alley' – is one of the most geographically perfect places in the world.

Hot air comes up from the south-east, around the Gulf of Mexico, while warm and sticky air comes up from the south-west. These masses of air come together around Oklahoma and Texas. The hot, dry air can over-run the warm, humid air and a lid forms, with the hotter air above.

'It's like a pressure cooker: the sticky air underneath wants to bubble up and burst into cloud but it can't because the lid is there,' says Ross Reynolds, an applied metereologist at Reading University. 'Occasionally, as the heating of the day progresses, the sticky air gets warm enough to pop through in a very dramatic way.'

The band between the sticky air and the hot air is called the 'dry line'. These start forming a day or so before the storms hit.

Weather forecasters look for these lines to warn people that they might be in for a tough few days.

For places like Oklahoma, May is the peak time for tornadoes. The temperature difference between the warm air at low levels and cold air above is at its greatest and a lot of churning is going on in the atmosphere.

How does a hurricane get its name?

US scientists responsible for tracking local hurricanes consult a master list of names decided upon by the World Meteorological Organisation.

There are lists for different parts of the world. 'In the Atlantic, they have six separate lists which all go alphabetically, alternate male and female names,' says Julian Hemming, tropical predictions scientist at the Met Office. 'At the beginning of each year, they start with the A storm and just work their way down.'

Every six years, the lists are recycled, except for the one or two names that are retired every year. 'If there is a very noteworthy hurricane, then they'll retire that name and replace it with another one,' says Hemming. A separate list for the north-east Pacific works on a similar basis to the Atlantic. Things are different for the north-western Pacific. 'It's made up of names which are, firstly, not an alphabetical list, and they're not proper names. They could be objects or even adjectives,' says Hemming.

In the south-western area of the Indian Ocean, there is a similar scheme to the Atlantic, which starts at A every year. But the names are not recycled: new ones are chosen every year.

Around Australasia, there are three regions of ocean with separate lists of names for each. At the beginning of a new season, the naming carries on where the last season left off.

'If storms do cross between regions there are varying practices on what happens,' says Hemming. The storm might even be renamed, particularly if it passes from the western side of Australia to the south-western Indian Ocean.

How do you stop a flash flood?

The answer is: you don't. You get out of the way. Flash floods are an accident of timing and topography. An inch of rain over one square mile adds up to 15 million gallons of water, for those comfortable with the old imperial measures: if the rain falls all at once and if this square mile tips straight into a narrow valley then nothing much is going to stop the cascade.

In Britain, flash floods usually cause only limited local embarrassment. In parts of the US, they are a permanent hazard. In June 1972, a thunderstorm in South Dakota led to a flash flood along Rapid Creek which killed 237 people. Even seemingly shallow flash floods are serious threats. Just a foot of water flowing sideways across a road can sweep an automobile off the tarmac and into a ditch.

Most flash floods are caused by slow-moving thunderstorms, or more occasionally heavy rains that follow hurricanes. A team at the University of Pittsburgh looked at thunderstorm-related deaths in the US between 1994 and 2000. Altogether, 1,442 people died. More than two-thirds of these were males, swept away by flash floods or struck by lightning. Most of those who died in flash floods were drivers; most of those electrocuted were playing sport. Flash floods tend to happen in arid, rock-strewn hilly regions off which water will flow at speed, after either an intense flow of rain or the breakup of an ice or debris jam.

The word 'flash' is not hyperbole: wadis and dried-up riverbeds become foaming torrents within minutes. Urban areas are increasingly at risk of local flash floods, simply because water cannot soak into tarmac or concrete. It runs off city streets two to six times faster than off grassland or scrub. In towns and cities, basements and viaducts can become death traps. The felling of forests, too, clears the way for sudden flooding. Just six inches of swiftly moving water can knock people off their feet. Flood waters can tear out great trees as if they were saplings, obliterate sturdy buildings and destroy bridges. Walls of water metres high can carry a lethal cargo of debris. If there is higher ground, head for it: on July 31, 1976, the Big Thompson river near Denver overflowed after a heavy storm. A wall of water six metres high roared down the Big Thompson Canyon. Holidaymakers were camping there at the time. Altogether 140 died in one night.

How do you build a hurricane-proof home?

Just make sure that everything is tied down firmly.

'If you can afford to build your building to withstand high winds, then there's almost no limit to the wind which you can design against,' says Brian Lee, professor of civil engineering at

Portsmouth University and chair of the UK advisory committee on natural disaster reduction.

When a building gets buffeted by winds, the most important force to consider is the one trying to tear off the roof. As the wind rushes by, the building will act like the wing of an aeroplane, where the air pressure above the roof will be lower than that below it. 'If you haven't tied the roof to the walls, then the roof comes off,' says Lee. Keeping the roof on in these circumstances involves more than just using a strong glue to stick on the tiles.

Lee says that, in a well-engineered house, the force acting on the roof will be spread to the foundations by means of a continuous path from one to the other. There should be no gaps, for example, between the walls and roof, or the walls and foundations.

The wind in a hurricane also fluctuates for up to 12 hours. 'So you've got this shaking effect,' says Lee. 'If you shake something, you can pull it apart more easily than if you just give it a steady pull. Part of the problem is that design rules don't take the shaking into account.'

The walls themselves also need to be strong. If debris punctures holes in the side of a house during a storm, the pressure inside can rise dramatically.

Will a meteorite land in your lounge?

It happens just a few times a year but there's no reason to think it's bizarre.

The Earth collects about 200 tonnes of stuff from space

every year according to Jay Tate, who runs the Spaceguard Centre in Wales. Most of it is in the form of cosmic dust, and burns up in the atmosphere without getting anywhere near the ground.

Monica Grady, a meteorite specialist at the Natural History Museum in London, says that more than 10,000 bits of rock over 10 g (which normally originate from the asteroid belt between Mars and Jupiter) end up landing on the Earth and they have a roughly equal chance of landing anywhere at all on the planet.

As we build more houses it gets increasingly likely that a meteorite will crash into one. 'Perhaps you could say once or twice a year, this sort of thing will occur,' says Tate.

Though meteorites land pretty much anywhere, where they are found is far from random. Most are found in desert regions such as the Sahara, or in Antarctica, where they can lie undisturbed.

Just as many meteorites land on a patch of, say, England as they do on an equivalent-sized patch of any desert, but finding them in such temperate regions is much more difficult. Put simply, meteorites that land here tend to get lost very easily. 'England is covered in trees, soil and landscape and if a meteorite falls in a bush, you've missed it,' says Grady. 'Also, in England we've got a more temperate climate so they rot down more quickly.'

What are my chances of being struck by lightning in bed?

Ask David Reilly of Ballymoney in Northern Ireland, who was watching television in bed when he suffered burns after lightning struck his house. The thunderbolt blew slates off the roof, demolished ceilings and – a doctor told Reilly – passed through his hand and out of his foot.

Terence Meaden, of the Tornado and Storm Research Organisation (Torro) at Oxford Brookes University, says: 'The lightning struck something on the roof and it ended up going into the ground. What route did it take on the way? That's the key.'

Electrical charge takes the path of least resistance, which could send it through cables in the wall and even the hot water system. A metal bedstead would help it enter Reilly's body.

Or it could have jumped through electrified air. 'Enormous current running through the electrical system might have been enough for a flash jump to him,' Meaden says. 'That's why we sometimes hear of 17 cows being killed instantaneously.'

Other explanations are possible. Peter van Doorn, a ball lightning expert at Torro, says: 'When ball lightning strikes a house the outer shell breaks away. The contents can go down the chimney and penetrate every room.'

Does removing road markings reduce accidents?

Probably not. Peter Chapman, assistant director of the Accident Research Unit at Nottingham University, is sceptical.

'Any change you make to the road environment will tend to have a positive effect in the short term,' he says. 'When you change, people notice the road is different, they're more vigilant and they're more alert.'

There are statistical anomalies to consider. 'Crash statistics for a particular road may increase dramatically in a calendar year for no other reason than statistical anomaly and then return to average levels,' says Andrew Meier of the Centre for Automotive Safety Research at the University of Adelaide.

'It is tempting to introduce countermeasures on roads where the crash rate is high but the rate may be about to fall regardless of the implemented measure. Alternatively, sites that are on low or average crash rates may show an increase despite the implementation of a countermeasure.'

The way to overcome anomalies is to carry out simultaneous research on many randomly selected sites.

Drivers often complain that the profusion of signs and markings on and around roads is distracting, and removing markings could therefore help. But there is little research to back this up.

Researchers at the University of Wisconsin, in fact, went the other way. In 1999, they painted extra chevrons on to the exit ramps of Interstate 94 at Milwaukee to see the effect on driving speed. 'The chevron markings reduce exit ramp speeds by creating the illusion that the vehicle is speeding and the road narrowing,' said the researchers. Average speeds fell significantly regardless of time – weekday or weekend, rush hour or not. They added that it was hard to generalise from one study but drew on Japanese research that also concluded that increased road markings reduced speeds.

A spokesperson for the Royal Society for the Prevention of Accidents said that, while any new method to reduce road accidents should be welcomed, highway authorities should be cautious about removing markings: 'Removing centre lines would not be appropriate on every road, and very often centre lines combined with side lines also encourage drivers to slow down because the road appears narrower.'

How do you make a city quiet?

Buildings are a big issue. 'Here, every apartment has air conditioning, offices are ventilated and the combined effect is a lot of white noise,' says Raj Patel, an expert on urban noise pollution working for Arup Acoustics. New York has stringent building regulations, but this adds to the problem. 'Because the code here in New York is so unrealistic, they just design whatever they want and the noise levels have just gone up and up and up,' says Joe Solway, also of Arup Acoustics.

New York also suffers from the 'canyon effect'. This happens when a lot of high-rises are built opposite each other and the sound from traffic and bars below simply bounces between them.

In Hong Kong, city planners have reduced noise levels by installing shields made of porous concrete along the sides of roads. Hong Kong has also put fins on skyscrapers, which deflect noise from the streets below away from the windows. Buildings can also be fitted with special sound-absorbing materials, although retrofitting Manhattan would not come cheap.

Why does moist toilet paper clog your plumbing?

It's obvious, really – moist toilet paper is designed to be stronger, so it takes longer to disintegrate. Which is why British Gas found that the number of toilets it has been called to unblock has jumped by 10%. British Gas blames premium toilet papers – often thicker and sometimes infused with lotions. These can take up to five days to break down in the sewage system.

A normal sheet of paper is made of a matrix of fibres that are held together by hydrogen bonds. When the paper gets wet, these hydrogen bonds break and the fibres start to unhitch. Things are different for moist papers.

'If you're applying a lotion to the sheet, then obviously you've got water in contact with those fibre bonds already,' says Richard Sexton of the Association of Manufacturers of Soft Tissue Paper. Without any additional chemicals, the lotion would just turn the loo roll into mush.

To prevent this, manufacturers coat the base paper with a resin that resists water. Or they might use synthetic fibres to make the paper, which can be bonded with heat. However, according to Sexton, the resins have a set lifespan and eventually do break down.

Science &
Technology

Could nanotechnology turn the world into grey goo?

Don't sell your house and flee to the hills just yet. If the notion of billions of miniature robots munching their way through our defenceless planet sounds like a great idea for a science fiction novel, then that's because it is. There may be risks associated with nanotechnology, but fears it will trigger a grey goo armageddon are fanciful.

The grey goo idea was first floated by Eric Drexler in 1986. He raised the prospect that 'nanobots' (created to build structures atom by atom) could produce endless copies of themselves, gradually tearing the world apart and leaving it a quagmire of grey goo.

The closest science fact comes to battalions of self-replicating devices is the attempt to direct organic molecules to assemble themselves into useful structures, for drug-delivery systems, say.

Many scientists do have concerns: some nanoparticles

could be poisonous. Their size makes them a potential health hazard for the lungs. Worrying for lab workers, but hardly the stuff of books and films.

How easy is it to identify someone from their DNA?

Relatively easy, with a decent sample of skin, hair, saliva, blood or bone and another to compare it with. And it helps, of course, if you have a rough idea of who you're trying to identify.

Humans are remarkably alike – one human differs from another by about 0.1% of their DNA – but this still adds up to about 3 million differences. In 1984, Alec (now Sir Alec) Jeffreys of Leicester University noticed that these tell-tale patterns in the human genetic code could serve as a kind of genetic fingerprint.

In 1985, this kind of DNA evidence solved a British immigration dispute by linking a child to a mother. In 1986, the world's first DNA-based manhunt unmasked a killer, Colin Pitchfork, who murdered two 15-year-old girls in Leicestershire.

Some DNA carried in the mitochondria of the cells is inherited only from the mother: this mtDNA has proved a powerful tool for solving ancient mysteries. It means that living descendants from the same maternal line can provide mtDNA, which can then be matched with that recovered from a long-dead body.

We know that the outlaw Jesse James did indeed die from a

single bullet wound in Missouri in 1882, because his sister's maternal great great grandson donated mtDNA that matched samples taken from unearthed remains.

DNA donated by a mother and sister helped US military researchers identify the unknown soldier of Vietnam at Arlington Cemetery. He turned out to be Michael Blassie, a pilot shot down in 1972.

DNA from handkerchiefs, toothbrushes and razor blades was used to confirm the identity of at least some human remains from the Twin Towers tragedy of September 11, 2001.

There are, however, two challenges: DNA breaks down, becoming harder to recover with time; and the world is smeared with it. Human beings leave behind invisible showers of skin, hair and body fluids wherever they go. So contamination is always a potential hazard for forensic researchers.

Will my email ever be hacker-proof?

You can never make something totally hacker-proof but, with a bit of quantum cryptography, you will always know if someone is trying to get at your messages.

Keeping private communications private relies on complex mathematical operations, which can all be cracked. You just need time. Try hacking a message encoded using quantum cryptography, though, and the laws of physics will instantly raise the alarm. Until now, messages encoded in this way were not strong enough to make much of a communication system. But scientists have succeeded in sending and receiving quantum-encrypted messages over 100 km – the furthest distance so far.

The technique is based on sending single particles of light, or photons, along an optical fibre. Each photon is encoded to represent a standard bit: 0 or 1. In a standard optical communication system, each bit is carried by a million photons. An eavesdropper could split off some photons – say a hundred or so – and determine the information they were carrying.

But because photons can't be split, the quantum technique is more secure. 'This stops a crude tapping-type attack where an eavesdropper tries to steal some of the photons,' says Andrew Shields, leader of Toshiba's quantum information group in Cambridge, which developed the system. 'In quantum cryptography, each bit is carried by a single photon so if you remove that photon then it's gone and the receiver never receives it.'

Quantum physics also tells us that you can't copy a single photon faithfully. 'If somebody tries to copy the signal they introduce changes, and those can be detected by the sender and the receiver,' says Shields.

One problem with the technique has been that photons are easily scattered out of the fibre. This reduces the signal which can get so low that it gets lost in the background noise.

'There's nothing we can do to reduce the scattering – that's fixed,' Shields says. 'We developed a very sensitive photon detector with very low noise counts and that meant we could tolerate a lower signal rate or have a longer fibre.'

How can you intercept other people's text messages?

The easiest way, of course, is simply to get hold of one of the phones used to send and receive the messages. But there are other ways, all of which need specialised equipment and a lot of preparation.

Text messages are sent over mobile phone networks in a similar way to voice conversations: the main difference is that the text is sent to a central processing centre, where it is stored until the recipient's phone is in reception range. According to a spokesperson for Vodafone, the message is then stored for up to 21 days (mainly to authenticate the sender and subsequently bill them) and then it is deleted. So anyone wanting illicitly to grab a text message between other people would have to do it during transmission. And that means being close to one of the participants.

'A common way of intercepting mobile calls is to fool the handset into thinking that your transmitter is its local base station,' says Anthony Constantinides, a communications professor at Imperial College London. 'The user then communicates through you, allowing you to listen in.' Potential hackers would have to be in the same cell, the geographic area served by a single base station, as the target.

Recovering deleted text messages from the phones themselves is also possible. 'It's the same sort of principle as when you delete a computer file,' says Fred Piper, director of the information security group at Royal Holloway, University of

London. 'It's still there until it's overwritten.' But this would require direct access to the phone or sim card.

Could we build a Star Trek phaser gun?

As Stephen Hawking once put it: 'The physics that underlies *Star Trek* is surely worth investigating.' Scientists say that in principle such a weapon might be possible, but that you'd never get it through the door. 'You need extremely high voltages and so it would be something the size of a car,' says Karl Krushelnick, head of the plasma physics group at Imperial College London.

In 1997 Eric Herr of the Californian company HSV Technologies patented a design using laser light. The lasers generate intense beams of ultraviolet light that create a path of ionised air between the weapon and the target, up to 100 m away. The air then conducts an electric current to cause muscle spasms or stun for a few seconds. Herr says he is developing a prototype device about the size of a suitcase.

How do you make bullet-proof glass?

Ask anyone in the industry and they will quickly correct you: it's bullet-'resistant' glass. 'There's no such thing as "proof" in this business,' says Phil Brown of Pilkington, the glass manufacturer. 'Proof suggests it will definitely stop anything fired at it. We tend to shy away from claims like that.'

Bullet-resistant glass is simple to make. Thin layers of a clear, tough plastic called polyvinyl butyral (PVB) are sandwiched between sheets of standard glass, and bonded together by heat. Alternating layers of plastic and glass are then built up. The more layers, the better the bullet-stopping power.

Bullets don't just bounce off it, of course. Fire a bullet at a pane of bullet-resistant glass and it will break the outer glass layer, and probably layers of glass deeper inside the pane. But the tough PVB sheets are designed to absorb its energy, preventing it from breaking through to the other side. At least, that is the idea.

Bullet-resistant glass is tested for strength using different calibres of guns. A shot from a modern rifle will need at least a 5 cm-thick pane to stop it, since the bullet speed is so high, typically around 820 m a second. Naturally, it takes less to stop a bullet from a handgun as the bullet travels much more slowly.

Some security companies offer 'one-way bullet-resistant glass', designed to stop incoming bullets, while giving the person on the inside the option to shoot back. The glass works by using a brittle glass layer and, again, a tough polymer layer. The brittle layer faces outward and shatters if a bullet is fired at it, spreading the force of the bullet over a large area, which is then absorbed by the tough layer behind. A bullet fired from the other side, however, can puncture the polymer layer easily before breaking the glass, only slowing the bullet slightly.

How does the army carry out controlled explosions?

The term 'controlled explosion', which is widely used in the press, actually covers a range of different operations. But the army often, perhaps surprisingly, uses water to carry them out.

There is understandable reluctance among defence ministry employees, army engineers and the companies that make bomb-disposal equipment to talk about how exactly they do it, so perhaps we should point out that everything that follows is from information already in the public domain.

For an example, take an abandoned car near an airport. Under such suspicious circumstances, the army's first task is to discover whether the car actually contains a bomb or not. Remote-controlled robots can be sent in carrying cameras and fitted with special tools and small explosive charges to open locked doors. In this case it's possible that the three controlled explosions carried out simply involved blowing open the locked doors.

If something suspicious is seen, then the level of alert – and the technology used – is quickly raised.

X-rays can help to judge whether an explosive device is present, and perhaps how sophisticated it is. But what really matters is the attempt to disable or disarm a bomb without accidentally triggering an explosion. This is where the water comes in. Very, very high powered jets of water.

'Water would be shot out of a cannon at very high velocity,' says one expert who works for a company that produces remote-controlled robots used in bomb disposal. 'It penetrates into the object, splatters everywhere and literally takes it apart. Hopefully the timing devices, the motion sensors and whatever may be in the set-up will not have time to make contact and ignite the explosion.'

Because the water used to shatter the bomb is cold, it is unlikely to ignite any explosive material. 'Also, being water it tends to short circuit all the circuits and wires in a fraction of a second,' he says.

In some cases the results can be astonishing. The British company PW Allen in Tewkesbury makes security equipment including the water 'disruptors' used in bomb disposal.

On its website it describes one piece of equipment called a 'disposable car boot disruptor', which can be placed on the ground beneath a vehicle believed to be carrying a large bomb. When activated, the device ejects water that physically tears the bomb apart with enough force 'to blow two filled aluminium beer kegs from the closed boot of a car to a height of approximately 10 metres', the site says.

How do you make a nuclear bomb?

In principle it's very easy – get a critical mass of radioactive material, sit back and watch the runaway nuclear reaction go. But luckily for us it's the first part – getting the radioactive material – that is the biggest stumbling block.

'You cannot make a nuclear bomb without fissile material,' says Andrew Furlong, of the Institute of Chemical Engineers. And for an average thermonuclear device, the necessary material is plutonium or enriched uranium.

Uranium, a naturally occurring heavy metal, comes as uranium 238 or 235. Both are radioactive and will decay into other elements, given time, but only the latter can be forcibly split when neutrons are fired at it. This is the basis of a nuclear bomb.

When an atom breaks apart, it gives out energy and more neutrons, which can then split other atoms. Get enough atoms splitting and you have the chain reaction needed for a bomb blast.

But natural uranium overwhelmingly consists of the 238 isotope, which bounces back any neutrons striking it – useless then for a bomb. To make a bomb, natural uranium needs to be treated to concentrate the 235 isotope within it.

And this is where the problems really begin. For every 25,000 tonnes of uranium ore, only 50 tonnes of metal are produced. Less than 1% of that is uranium 235. No standard extraction method will separate the two isotopes because they are chemically identical.

Instead, the uranium is reacted with fluorine, heated until it becomes a gas and then decanted through several thousand fine porous barriers. This partially separates the uranium into two types. One is heavily uranium 235, and called 'enriched', while the rest is the controversial 'depleted' uranium used to make conventional weapons.

To make a nuclear reactor, the uranium needs to be enriched so that 20% of it is uranium 235. For nuclear bombs, that figure needs to be nearer 80 or 90%. Get around 50 kg of this enriched uranium – the critical mass – and you have a bomb. Any less and the chain reaction would not cause an explosion.

You could use plutonium instead. According to Keith Barnham, a physicist at Imperial College London, this is the preferred material because it makes much lighter weapons that can be mounted on to missiles.

Plutonium is produced as a by-product in nuclear reactors and only around 10 kg is needed for a bomb. An average power plant needs about a year to produce enough and expensive reprocessing facilities are required to extract the plutonium from the fuel.

The bomb will explode once the critical mass of uranium or plutonium is brought together. So, to begin with, and to make sure that it doesn't explode in the hands of its owners, the bomb needs to keep the metal separated into two or more parts. When the weapon is in place and ready to go off, these sub-critical masses need only be thrown together – and this can be done with conventional explosives.

The chain reaction, explosion and familiar mushroom cloud then take care of themselves.

How do you tap a phone?

Shockingly easily. If you thought that your phone conversations were secure, think again. Covertly listening in on phone calls is a doddle if you know how.

'It's almost as easy as plugging in something to an electrical outlet, that's the scary part about it,' says Grant Haber, president of American Innovations, manufacturers of covert and counter-surveillance equipment.

Telephones are very simple devices. A microphone converts your voice into electrical pulses, which are then relayed through a wire out of your home and through several telephone exchanges on the way to the person you are speaking to. At any point in this line, somebody can simply attach a device to the wires which will convert the electrical information back into sound.

At its simplest, this device can be another telephone. More sophisticated bugs convert the electrical current into radio waves and transmit the information to a receiver – usually a van parked somewhere near the transmitter.

Mobile phones are more difficult to intercept, but it's still relatively easy to do. For around £250 you can buy equipment that allows you to tune in to any calls going on in your area.

'A normal mobile telephone isn't actually secure,' says Anthony Constantinides, professor of communications and signal processing at Imperial College. 'There is no encoding procedure that actually secures such conversations.'

The digital mobiles in use today are encrypted but the codes all conform to international standards so that phones can work overseas. 'Anybody can undo the encoding,' concludes Constantinides.

Another common way of intercepting mobile calls is to fool the handset into thinking that your transmitter is its local base station. The user then communicates through you, allowing you to listen in. The only drawback for a potential hacker is that they have to be in the same cell – the geographic area served by a single base station – as you listen in.

To get around this, satellites can be used to hone in on any particular spot. 'Satellites can transmit information much in the same way as a base station,' says Constantinides.

And there are easier ways for governments, for example, to

listen in to phone conversations – such as asking the phone operator to patch them in.

Can iPods make you hallucinate?

If you like your answers based on proof, then this particular one has to be a firm 'no'. But the issue has been raised by Victor Aziz, a psychiatrist at Whitchurch Hospital in Cardiff and expert in so-called musical hallucinations.

Just like the more familiar visual variants, musical hallucinations strike suddenly. 'People will all of a sudden start hearing a song, such as "Yes, we have no bananas",' says Aziz.

Musical hallucinations are rare and usually linked to some kind of abnormal behaviour in the brain, be it a psychiatric condition, epilepsy or a tumour. But Aziz says people are more likely to experience them if they go from hearing a lot of music to a quiet place in which their brains receive little auditory stimulus.

Traditionally, scientists thought that hallucinations were more common among those who listened to a lot of music in childhood, but Aziz found that many of his patients were hearing more recent songs.

Aziz believes that in the iPod age, the increase in the amount of music we are exposed to will make hallucinations more common. 'We are now exposed to a barrage of music and it seems that we might well see more cases of this in the future,' he says. 'We'll only know if we test people in twenty years' time,' he added.

Ironically, iPods and Walkmans are used by many patients

who experience intrusive musical hallucinations, says Peter Woodruff, a psychiatrist at Sheffield University. 'What they find is that by playing real music, it competes with the hallucination and suppresses it,' he says.

Some auditory hallucinations are normal. On falling asleep and waking up, it is fairly common to think you've heard your name called, or less specific noises, Woodruff says. 'It's when they happen outside these times that you want to see a doctor.'

Brain scans of people experiencing musical hallucinations show that neural activity is identical to the state of really hearing the music. 'It's not like having a tune going around in your head,' said Adrian Rees, an expert in auditory neurology at Newcastle University. 'This is something you can't turn off or change to another record.'

Have you got BlackBerry thumb?

If you are a text message addict and own a BlackBerry, the hand-held gadget with email, text, pager and mobile phone, then you may have experienced BlackBerry thumb. Reports say thousands of BlackBerry owners have been turning up at doctors' surgeries complaining of aching thumbs from using the device's tiny keyboard. Frantic texting on ordinary mobile phones may also leave people vulnerable to the condition. So is the condition likely to reach epidemic proportions?

'At the moment no one knows, but a big uncontrolled experiment is happening right now because so many people send texts,' says Roger Haslam, a professor of ergonomics at

Loughborough University. 'The number of young people who intensively use mobile phones is particularly alarming.'

Thumbs were never designed for texting, so it is not surprising that they complain when asked to tap out messages all day long. 'The most obvious problem is tendon injuries,' says Haslam. If thumbs are repeatedly bending then the tendons begin to rub as they are stretched over the knuckle joint. Eventually this leads to swelling and pain. 'Tendon injuries can be difficult to get rid of and once you have had one then you are often susceptible to it happening again.'

Alternatively, BlackBerry thumb sufferers may be showing the early stages of osteoarthritis. 'The thumb has more than two planes of movement, allowing it to flex and rotate. This means that some people develop pain at the base of their thumb associated with osteoarthritis,' says Sean Hughes, professor of orthopaedic surgery at Imperial College London.

What can you do to avoid it? 'Take lots of breaks and stretch the fingers regularly,' Haslam says. In addition it may be worth tailoring your phone to your hand. 'Find a phone that feels comfortable and listen to your body. If you start to get aches and pains then do something about it early on.'

Why do wind turbines confuse military radar?

Put simply, one piece of fast-moving metal looks pretty much like another to a radar operator, whether it's the rotating blades of a wind turbine or the approach of an enemy aircraft. Which is why the Ministry of Defence takes such an interest in where green energy developers intend to erect them. 'There are genuine concerns over how wind turbines can interfere with our radar systems,' says the MoD.

The problems start with the fact that wind turbines are very large, made of metal and have sharp edges. Sound familiar? They would if you were sat at a radar listening for returned 'pings' bouncing off aircraft – in fact they might sound exactly like a jumbo jet. Hence civil airport authorities and air traffic controllers have a problem with wind farms, too.

The rotating turbine blades fool techniques used to filter out tall buildings, trees and other stationary objects. And because different blades can be picked out during different radar sweeps, banks of turbines appear as a confusing, twinkling mass on screens that can make genuine targets difficult to pick out. There are even concerns that turbines cast a radar shadow behind them, within which enemy planes would be invisible, though measurements indicate that it would last for only a few hundred metres and would hide only very small objects.

The government – which has promised to generate 10% of electricity from renewable sources by 2010 – and industry are investigating a number of solutions.

Software fixes that help radar systems filter out signals from wind farms have been developed, though these work better with offshore wind farms surrounded by lots of flat sea. Another option is to redesign the turbines and the way they are arranged so they better blend in with terrain.

The military research company Qinetiq is using stealth bomber technology to build turbine blades that don't show up on radar screens. 'We're looking to change the properties of parts of the material structure to reduce the amount of reflection,' says Andy Beck, a radar expert with Qinctiq. Making the turbine blades from different layers the right thickness can bounce back signals that neatly cancel out the arriving pings. And honeycomb-style foam can absorb enough of the incoming radar energy to send very little back.

What would it take to decontaminate your home?

The simple answer is lots of soap, water, mopping, vacuuming and elbow grease. But for a chemical, biological, radiological or nuclear contamination, some extra technology may be required. According to the government's new booklet that advises on how to prepare for an emergency, we may have to wait for the emergency services to decontaminate buildings before we can go home. So what might the emergency services be doing and will we want to go back home afterwards?

One technique, used on chemical and biological contaminants, is to pump reactive gases and vapours into the building, to react with the contaminants and mop them up. Reactive gases and vapours (such as chlorine dioxide or hydrogen peroxide) permeate into porous materials, like soft furnishings, carpets and wallpaper, react with the contaminants and then diffuse out again. For more targeted and precise cleaning operations, peelable coatings can be applied to a contaminated surface. These sticky strips are coated in a polymer that binds with the contaminant and are useful for awkward areas.

Although these methods will make your house sparkling

clean, they can also leave a rather unpleasant smell. In Florida, the *National Enquirer*'s AMI building (contaminated with a letter containing anthrax in September 2001) was decontaminated using a strong disinfectant called para-formaldehyde. The building is still vacant as a result. 'The problem with formaldehyde is that it smells so badly, no one wants to use the building afterwards and it's so corrosive, it'll strip the paint off your walls,' says Ray Zilinskas from the Centre for Non-Proliferation Studies in Monterrey, California.

For radiological and nuclear contamination, high-pressure hosing, steam-cleaning and extensive vacuuming may be the way forward. Of course, all the debris from this intensive cleaning has to be collected and disposed of in a site for hazardous waste. Organic solvents with low boiling points (such as acetone) also come in handy. They can be heated until they vaporise and then circulated throughout the house. The vapours permeate porous materials, where they condense and dissolve the contaminant, before diffusing.

These techniques are all costly: around $800 million has been spent on decontaminating the 23 buildings in the USA that were contaminated with anthrax. Unless your house is very special, it may just be demolished. 'The cost of decontamination can become very expensive. It may very well be cheaper to raze the building and start all over again,' says Jerry Loeb, also from the Centre for Non-Proliferation Studies.

How big could we build an aircraft?

Bigger than the new Airbus A380, that's for sure. And the

world's largest passenger jet is some way from breaking any records.

Its 79.8 m wingspan is no coincidence. Aircraft wanting to use boarding gates and taxiways at the world's airports need to operate inside an 80 m box.

The biggest aircraft in the sky is the Russian Antonov 225 cargo plane, a full 10 m longer than the Airbus, with a wingspan more than 88 m. Bigger still, Howard Hughes lifted off in his *Spruce Goose* flying boat, which had wings an incredible 98 m across. Debate still rages about whether his brief, low-level test flight in 1947 really counts.

As a plane gets bigger and carries more passengers, it weighs more. Heavier planes need more lift and so bigger wings. Kenji Takeda, an engineer with the aerodynamics and flight mechanics research group at the University of Southampton, says conventional wings can only grow so far before they are unable to safely support their own weight.

'With the A380 I know they had some problems in terms of how big a bit of aluminium they could get,' Takeda adds. Some of the pieces have to be made from the same original ingot for structural reasons.

He says the new Airbus is probably at the peak of the trade-off between size and efficiency. Future designs could see a return to biplanes or a new 'flying wing' concept to generate yet more lift.

The European manufacturer has stressed the plane's green credentials as the first long-haul aircraft to consume less than three litres of fuel per passenger over 100 km – comparable to an economical family car.

Why do aircraft wings now go up at the ends?

Aircraft manufacturers claim that winglets, as they are known, cut drag and boost fuel efficiency by up to 5%. Though they have been reported as a new trend, they have been around longer than you may realise.

NASA first realised their aerodynamic benefits in the 1970s and, as the price of aviation fuel has soared in recent years, winglets have become the latest must-have in the skies. Boeing says requests for winglets on its 737s are up from 10% in 2001 to 50% this year.

Philip Butterworth-Hayes, editor of *Jane's Aircraft Component Manufacturers*, says: 'I reckon you're only looking at 1–2% increase [in fuel efficiency] using a winglet but that is really quite significant. I reckon they will soon be on every airplane.'

British Airways has winglets on 57 jumbo jets and 66 shorter-haul Airbuses, which were in place when the planes arrived from the manufacturers. Rival operators have invested in kits that equip their older aircraft with the ski-shaped ends. The fins can reach up to 4 m above the wing and work by evening out the air flow around the tips.

'It's well-known that modifying the wingtip flow is important,' says Kenji Takeda, an engineer with the aerodynamics and flight mechanics research group at the University of Southampton. Soaring birds such as eagles have strong feathers that flip up at the wingtips to reduce drag and give the birds more lift. 'Nature, as always, has sussed it out first,' Takeda adds.

Winglets could bring other environmental benefits besides saving fuel. The altered air flow around the wingtips also reduces the formation of contrails, wispy streaks of cloud left behind when water vapour condenses around particles of pollution in engine exhaust fumes.

How contrails could influence global climate is still debated, though some scientists say they promote the formation of long-lasting cirrus clouds, which help to trap heat at the Earth's surface. Last year, NASA scientists said an increase in cirrus cloud cover over the US of 1% a decade since 1975 was down to air traffic.

Why do people keep on looking for ever bigger prime numbers?

Trying to find big prime numbers is a useful way of testing computers, and very big prime numbers can be used to help encrypt electronic information. But there's also the geek factor: big prime numbers are the sort of thing amateur mathematicians become obsessed by.

Prime numbers, numbers that are only divisible by themselves and 1, are a mathematical oddity. They appear seemingly at random along the number line.

Finding small ones (1, 3, 5, 7 etc.) is obviously easy – just divide each candidate number by all the smaller numbers and see if any of them go in a whole number of times.

As the numbers get bigger, however, this becomes unfeasible and you need some serious computing power. Prime numbers are widely used to produce encryption codes on the internet – when you submit your credit-card details to a website, for example, it will be encoded using methods that use prime numbers.

'The security of those systems are based on the fact that it is very hard to factorise integers into primes,' says Alexei Skorobogatov, a mathematician at Imperial College London. But to make that work effectively, you need big prime numbers.

For professional mathematicians, the allure of primes lies in the fact that they are seen as the building blocks of numbers.

'Every whole number is a product of prime numbers,' says David Solomon, a mathematician at King's College London. 'That is like the signature of the number.'

Primes, and number systems based on them, are used extensively in theoretical mathematics as tools to solve complex equations; when Andrew Wiles, a professor of mathematics at Princeton University, solved Fermat's last theorem a decade ago, he worked with new mathematical techniques that use number systems based on prime numbers.

But academics themselves tend to shy away from the search for ever bigger numbers.

'Mathematicians don't, generally speaking, go around looking for prime numbers. The main reason is that we know there's infinitely many prime numbers, so you're never going to get to the end of the list,' Solomon says. Instead, he says, they concentrate on more general questions such as trying to work out if there are indeed any patterns in how the numbers appear.

Why is artificial blood so hard to make?

Because blood is made of many complex parts that serve specific functions: it's tough to reproduce each one properly.

But Eishun Tsuchida, a biochemist at Waseda University in Tokyo, says he's solved the problem. Using yeast to artificially manufacture human blood proteins, he claims to have produced the world's first entirely synthetic red blood cells.

Hospitals always want as much blood as possible but there are risks that donations could be infected with CJD, hepatitis viruses or even HIV. 'For a long time people have been trying to make replacements,' says Sarah Middleton, chief executive of Haemostatix, a company that makes components for artificial blood.

So far, biotechnologists have looked at just one part of the puzzle. 'When you need blood you need it for a particular purpose. You either need it to make a blood clot or you need it because you need more oxygen or fluid,' says Middleton.

Each bit of the blood has its own problems. 'The difficulty in making the red cell component is that you can't really make cells that easily,' says Middleton. People have tried to make the oxygen-carrying part of blood, a molecule called haemoglobin, which is the main component of red cells.

These are made up from proteins called globins and haeme, a small molecule that actually carries the oxygen. 'The reason that haemoglobin is in a red cell is because there are certain moderating influences within the red cell that make sure that that happens correctly,' says Middleton. 'The other thing is that haemoglobin is small: if you just have a single globin with haeme on it and you were to inject that into somebody, it was would [go straight] through the kidneys and have no half life of circulation.'

Producing globins on a large scale is also difficult. At the moment, they are made by inserting a gene into yeast and allowing the organism to make them slowly but surely. But it would be difficult to make globins on an industrial scale this way.

And where do you get the haeme from? The US firm Biopure has successfully made artificial blood from a polymer of artificial haemoglobin molecules and its makers claim it is more efficient than real blood because it absorbs and releases oxygen three times faster. It is also less viscous than real blood, so can flow past obstructions that could block normal red blood cells. But its haeme is extracted from cows' blood, and is unlikely to become popular in places such as Britain, where the fear of mad cow disease is still fresh.

How do you win a Nobel prize?

Who better to enlighten us than three of Britain's own Nobel laureates, Tim Hunt, Harry Kroto and John Walker?

'As with all human affairs, it pays to be incredibly clever, incredibly hard working and incredibly lucky. But to be frank I'd put the emphasis on the last one,' says Tim Hunt of Cancer Research UK, who with Paul Nurse won the Nobel prize for physiology or medicine in 2001.

Sadly, Lady Luck, and the fact that she might well forget to look your way, is something any wannabe Nobel prizewinner has just got to accept. But setting luck aside, there are ways of stacking the odds of landing a Nobel in your favour. One way is to take a chance on finding something no one else is even close to discovering.

'These Nobel people are really keen on discovery with a capital D and with pioneers, people who really opened a door on something when nobody even realised it was there,' says Hunt.

If taking the chance leads you to something interesting, it could give you a far better chance of winning a Nobel than jumping on the latest bandwagon. 'You can be a brilliant scientist, be fantastically effective and have a huge team of people churning out results and never get within a million miles of winning a Nobel prize because you are treading a well-worn path. Everything you find out is more or less expected,' says Hunt.

The best bet, according to Hunt, is to find just the right kind of experiment. 'It's no good trying to understand consciousness, you'll just flail away getting nowhere, and it's no good wanting to know how your left foot goes in front of your right because anybody can do that. But somewhere in between is that land where you might make a really big difference and discover something really important. That's where you want to focus your effort,' he says.

There is a risk in pursuing obscure scientific problems that no one else is interested in though. You could end up in an academic wasteland so bereft of interest that you fail to win grants and slowly find yourself unemployable. The key, whether by intention or accident, is to do the right experiment at the right time, says Harry Kroto of the University of Sussex, who won the Nobel prize for chemistry in 1996. 'The most important experiments are the ones where you really don't know beforehand where you are going or what you are going to find,' he says.

At the very minimum, you need to be doing science because you feel a need to crack whatever problem it is you are studying. 'If it's hard work and you enjoy it, that's a good start,' says Kroto. 'The only recipe I have for my research is that if it is interesting to me I do it.'

John Walker, director of the Medical Research Council's Dunn human nutrition unit in Cambridge, and winner of the Nobel prize for chemistry in 1997, says that one of the worst mistakes you can make is to want a Nobel prize too much. 'I've met several people who set themselves the task of winning a Nobel prize and most of them ended up very disappointed,' he says. 'I think it's dangerous to assume you can win a Nobel prize because many people do good enough work, but for whatever reason they don't actually win the prize. At the end of the day, it's in the hands of the Nobel academies in Sweden and it all comes down to what they perceive as worthy of it.'

Can science prove the existence of God?

No. Any cosmologist would have a hard time 'proving' the existence of anything that exists before time or beyond space. An evolutionary biologist is in no position to demonstrate precisely how life began, simply what paths it took after it did begin.

Many scientists do believe in a personal God, but not because they have scientific evidence for Him. Some state that they are in no position to say whether God does or doesn't exist. Isaac Newton and John Ray embarked on a study of the cosmos and of life on Earth, because they believed it would reveal God's handiwork. Gradually, even the most devout began to accept that science showed no such thing.

'I flatly reject the argument that the origin of life was some sort of miracle,' says Paul Davies, author of a book about cosmology called *God and the New Physics*. 'To be sure, we don't yet know how it happened, but that doesn't mean a cosmic magician is needed to prod atoms around.'

What is biodynamic farming?

It is about burying cow horns full of manure and planting crops according to signs of the zodiac. Reports say the Prince of Wales has decided to experiment with some of the principles of biodynamics, a type of agriculture founded by an Austrian philosopher, Rudolf Steiner, in the early 20th century.

'Steiner said you should treat the farm as an entity and know that whatever you do on one part of the farm affects it elsewhere,' says Alan Brockman, a Kent-based farmer with 40 years' experience of biodynamics.

Brockman says biodynamics means taking account of natural cycles when farming. 'Say we want to plant carrots. We'll pick out a constellation, such as Virgo, Capricorn or Taurus and plant them when the moon is moving through them. Those constellations all stimulate root development,' he says.

But Geoff Squire at the Scottish Crop Research Institute says that, while seed germination depends on differences in temperature, sunlight and moisture, there's no evidence that the moon makes any difference. As one scientist puts it: 'Biodynamics? It's kind of an occult-based farming system.'

Isn't it dangerous to use a mobile phone on a plane and in hospitals?

Apparently not. For many years, mobile users have diligently switched off their phones after being told that sensitive electronic equipment on jets or in hospitals might malfunction under the influence of the microwave radiation emitted by the phone.

But it seems it may all have been unnecessary. 'They were erring on the side of caution more than anything else,' says John Pollard, a lecturer in indoor personal networks at University College London, about attitudes to mobile use on jets and in hospitals.

Indeed, research carried out by the Medical Devices Agency in 1997 showed that outside a metre or so of sensitive medical apparatus, there seemed to be no danger at all. 'I can imagine that aircraft devices are much less susceptible,' Pollard says.

A mobile phone base station in a plane would work by communicating directly with satellites if the plane is over water, or with normal base stations if they are flying over land.

Airbus wants to introduce the technology on short-haul

flights in the first instance, to avoid the possibility of passengers who might be asleep on long-haul being disturbed.

As for hospitals, 'what I'm told is that doctors ignore the ban anyway,' says Pollard. 'They've all got mobile phones and they use them willy nilly.

Are fireworks getting better?

You bet. Digital simulations, electronic fuses, smokeless compressed-air propellant and even rockets carrying computer chips have all fizzed over the horizon in recent years, although the chances are that your local November 5 display will rely on more conventional methods.

'The biggest difference is in the machines used to fire the displays,' says John Bush of the British company Millennium Fireworks. Large displays are planned using computer simulations, which allow the bangs and whistles to be tightly coordinated with music and lights, if required.

And forget the well-wrapped-up men shuffling around with lit tapers – the firing of the fireworks in big displays is done digitally too, often using signals sent from a computer.

When it comes to the fireworks themselves, not much has changed in years. They still work on the idea that different metals give off different wavelengths of light when they burn to produce colours, and most use the same delay fuses to place the bangs in the right place. 'There's not a lot of difference, although colours have changed slightly and there's a slightly wider palette than there was 20 years ago,' says Tom Smith, who ran the London millennium firework display.

Fancier fireworks are available, but at a price. To reduce the smoky haze left by its nightly displays, Disney has experimented with launch tubes fired by compressed air, and some manufacturers even offer computer chips inside the explosive shells, which they say allow such accurate timing of explosions that words can be spelled out. 'It's possible but it's incredibly time consuming and unbelievably expensive,' says the Reverend Ronald Lancaster, a renowned firework expert.

And the million-dollar question, is there a difference between an 'ooh' and an 'ah'? Bush says there is. Pretty displays such as fountains of sparks falling from bridges tend to draw the latter, while punters save the former for sheer extravagance. 'That's an ooh because you're just firing so much money in such a short space of time,' he says.

Who would you look like after a face transplant?

No one in particular. When a donor's face is spread over a recipient's skull and facial muscles, the effect is a hybrid that doesn't look like either person.

Unlike in the 1997 film *Face/Off*, in which Nicolas Cage and John Travolta have their faces switched, the recipient doesn't end up looking exactly like the donor.

'It's a big question, because donors' families don't want to walk down the street and see the face of their loved one,' says Peter Butler, a surgeon who is investigating the issues surrounding face transplants at the Royal Free Hospital in London.

'What happens is you get the skin tone and texture and the

hair colour transferred, but the bone structure dictates most of what the face will look like. The end result is they look like someone different,' he adds.

Anyone who has a face transplant in the future will need to have counselling beforehand to prepare themselves for how they will look afterwards.

'One of the problems is that there is now a lot of high expectation around about face transplants,' says Butler. 'It's important to remember these people will have severe facial injuries and the transplant will be an improvement. But it is unlikely they will look completely normal. There will be scarring and tissue damage that will affect the final outcome,' he says.

Why is it dangerous to clone humans?

The high failure rates reported for cloning animals is an indicator. According to Wolf Reik, of the Babraham Institute, Cambridge, around 99% of clones die in the womb or suffer genetic abnormalities.

But what goes wrong? The problem is that the DNA used to make the clone is taken from cells that aren't meant to create embryos. When a cell matures and turns into a particular cell type, such as skin, it programmes its own DNA to express the right genes at the right time to become, and remain, a skin cell. This is done in two ways. Firstly, chemical compounds are tagged on to the central protein thread (chromatin) that DNA is wrapped around. Second, compounds called methyl groups latch on to specific genes, governing when and if each gene is switched on or off. The way the DNA is programmed is different for every tissue type.

It's no surprise then that skin cell DNA can lead to appalling defects if used to grow an embryo. 'You get the wrong pattern of gene activity during development, so the clone dies early in the womb or has developmental abnormalities when it is born,' says Reik.

But the very fact that some cloned animals are born, at least superficially, quite healthy, suggests that every now and again, the DNA is able to 'forget' what kind of cell it used to be and apply itself to making an embryo. Scientists know that it is chemicals in the body of the hollowed-out egg that help reprogramme the DNA, but quite how remains a mystery.

Harry Griffin, deputy director of the Roslin Institute, which gave us Dolly the sheep, says claims that genetic abnormalities produced by the cloning process can be detected before birth are nonsense. 'There's no way you could pick up some of these subtle, but life-threatening defects,' he says.

There may be another barrier to human cloning. Some studies have suggested that clones born successfully have a biological age the same as the animal that donated the DNA. 'It could mean you age far more quickly,' says Reik.

Has stem cell research been over-hyped?

Lord Winston, fertility expert at Hammersmith Hospital in London, thinks so. He singled out claims surrounding research into embryonic stem cells as being particularly over-blown. The danger, he said, was that hype could lead to public expectation becoming unrealistically high, setting up an inevitably painful fall when scientists fail to come up with breakthroughs in the near future.

Richard Ashcroft, a bioethicist at Imperial College London, says the stem cell hype might not be as bad as Winston makes out. 'To build public support for what they're doing, scientists are always going to say there is the prospect to cure all these horrible diseases, but they're cautious for the most part in saying when those cures will arrive.'

But stem cell researchers, especially those working on embryonic stem cells, which must be harvested from early-stage human embryos, may be more prone to hyping their work than others, adds Ashcroft. Because embryonic stem cells are deeply frowned on by many religious groups and others who disagree with the use of human embryos in research, scientists are under greater pressure to extol the potential of their work.

Space & Time

Are we ready to send humans to Mars?

Nowhere near it. But that hasn't stopped the US promising to have humans living on Mars 24.7/7 (the Martian day is 39 minutes longer than ours). All that needs to be sorted out are the scientific and technical problems in getting there and back in one piece without going hungry, thirsty, dying of radiation sickness, going mad in such a small container with so few people, wasting away or suffocating.

The trip to Mars is at least a six-month jaunt either way, but once there, any Mars explorers will have to keep themselves busy for 18 months before Earth and Mars are again in a suitable position to make the journey home. Food and water can be sent ahead on separate rockets, but six months of supplies will still be needed for the trip out. Since any mission would take between six and eight astronauts, to ensure enough engineers, geologists and medics are on hand, that means a lot of food.

But food may be a minor worry for any Mars-bound astronaut. 'We still don't really understand how humans cope with zero gravity,' says David Williams at NASA's Goddard Spaceflight Centre in Maryland. 'Astronauts that have been in space for a year come back and they can't walk. If you get to Mars and you can't even walk when you get there, what's going to happen? At least when they land on Earth, there are medical facilities waiting for them,' he says.

Astronauts struggle to walk after being in zero gravity because without the pull of gravity, their muscles waste away. Their hearts also weaken as it's easier to pump blood around the body. So even if humans do get to Mars, they could be physical wrecks on arrival.

Have we infected Mars?

Robotic instruments sent to search for life on another planet need to be scrupulously life-free themselves – NASA takes the business of 'planetary protection' very seriously and insists on ultrasterile conditions of manufacture. But microbes are very small, difficult to detect and fiendishly well equipped for survival.

A book, *Out of Eden*, by Alan Burdick, says that a microbe called *Bacillus Safensis* – said to have evolved to survive within Nasa's Jet Propulsion Laboratory's spacecraft assembly facility – is highly resistant to gamma and ultraviolet radiation and can draw energy from the ions of trace metals such as aluminium and titanium.

Its discoverer, Kasthuri Venkateswaran, thinks the bacillus

could have travelled to Mars, and survived in the rovers. If so, then some future mission searching for extraterrestrial life could reach Mars, and, embarrassingly, discover a colony of microscopic Earthlings.

But scientists have already discovered Earthlings on another heavenly body. A camera in a *Surveyor* probe sent to the moon was retrieved by the crew of *Apollo 12* almost three years later and shipped back to Earth again. Within the camera were a colony of *Streptococcus mitis*, a tiny microbe that had inadvertently stowed away. The *Streptococcus* survived three years in a vacuum, experiencing extremes of heat and cold, without food or water, bombarded by lethal radiation. When it got back to the home planet it revived, and began to multiply.

So, if bacteria can survive in space, perhaps they already have? According to one estimate, Martian rocks hit the Earth as meteorites at the rate of half a tonne a year. The British-born cosmologist Paul Davies – now at the Australian centre for astrobiology at Macquarie University – has argued for years that life could, plausibly, have begun on Mars when the planet was wet and warm, and been exported to Earth inside a lump of Martian shrapnel. In which case, could the colonists have survived, mutated, evolved into complex organisms and 3 billion years later started sending probes to Mars? In other words, did Mars infect us?

How much training do you need to go into space?

Depends what you want to do. If you're planning to take a ride on a tourist vessel such as Sir Richard Branson's proposed

Virgin Galactic service, the plans are that you'll need a week, as opposed to the years it takes to train as an astronaut.

What that training will involve is unclear at present but David Ashford, director of Bristol Spaceplanes, reckons that a week is overkill.

'You could do it in a day,' he says. 'All you're doing is sitting in a seat and you're strapped in – you've got no control over anything. What Virgin Galactic are talking about is a very brief whip up to space and back, being out of the atmosphere for maybe a few minutes.'

The training you need will mainly get you used to the g forces experienced during the trip.

'When you come back in from space, you pull out of a [steep] dive,' says Ashford. 'You're talking about four or five g for 10–20 seconds. That's quite a lot. You probably have to go in a centrifuge [in training] to check out you're OK.'

Prospective passengers will also need a medical exam, but Ashford says this will probably be straightforward. In any case, a lot of the training will be just a precaution.

'Because this is a new industry, the authorities will start off being cautious and only take fit people and give them a lot more training than they need,' says Ashford. 'We'll find out by trial and error what training, what medical tests are needed.'

Ashford says that if the Virgin plan is successful, it could trigger the far more tantalising idea of developing commercial spacecraft that can go into orbit, and potentially send people into space to stay in hotels there. In this case, passengers would need lots more training. They would be wearing pressure suits (so they'd need to know how to operate those) and would need to be happy with weightlessness by training in aeroplanes on parabolic flights – the so-called 'vomit comets'.

Ashford says: 'In 15 years' time there will be a million people a year going to space hotels.'

What will NASA do on the moon?

The ultimate plan is to build a base and keep astronauts there permanently: a step on from the permanent presence in space afforded by the International Space Station and a practice run for any future adventures to Mars.

NASA plans to get humans back on the moon by 2018, almost 50 years after the *Apollo* astronauts last walked on it.

Exactly what form the rudimentary moon base will take is still in the earliest stages of planning, but NASA did give some clues. Because taking equipment up into space is so prohibitively expensive, scientists want the astronauts to build as much as they can with materials on the moon itself. The priority for the next decade in space technology, then, will be producing mini-factories which will be able to process the raw ingredients available on the moon.

The day after the space agency unveiled its plans, it announced a $250,000 prize for scientists and inventors to come up with a machine that can excavate the most soil and deliver it to a collector. The digger will be used by robots. They will be dispatched to the dark craters at the moon's poles to find out if there is water ice there, a source of rocket fuel, oxygen and water to keep crew and equipment going.

The diggers will also mine ilmenite, a mineral from which

astronauts can extract oxygen, hydrogen and helium. This could produce air and water, while the flammable gases could be burned to generate electricity.

NASA said the lunar base would provide a 'huge head start in getting to Mars. A lunar outpost just three days away from Earth will give us needed practice of "living off the land" away from our planet, before making the longer trek to Mars.'

What use is the transit of Venus?

As an event, it's pretty unique. But historically the transit of Venus has been much more than a mere movement of heavenly bodies. In 1639, astronomer Jeremiah Horrocks unwittingly inflated the size of the universe by using the transit to measure the astronomer's favourite yardstick, the astronomical unit – the distance from the Earth to the sun. He timed how long Venus took to move across the sun from two different positions, then used trigonometry to work out how far away the sun must be. The answer he arrived at, 90 million km, was nearly 10 times greater than scientists had thought. 'With that one calculation, he expanded the solar system,' says Robert Walsh, an astrophysicist at the University of Central Lancashire, Preston. More recent measurements, achieved by bouncing radio waves off the sun and timing how long it takes them to return, have highlighted inaccuracies in Horrocks's method: an astronomical unit is now known to be around 150 million km.

What is Galileo really for?

Galileo is Europe's planned rival to the ubiquitous, US-owned GPS satellite positioning system. Publicly, it has been touted as a purely civilian system, but some say that it has clear military uses too.

According to Dominique Detain, a spokesman for the European Space Agency (ESA), which is building the system, Galileo has been designed solely for civilian uses, such as tracking ships and delivery trucks.

That doesn't mean it can't be used by the military if nation states decide they want to. Like GPS, Galileo will have publicly available signals and more accurate encrypted signals only available to governments.

According to one expert close to the US/European negotiations over Galileo, it is referred to only as a civilian system for political reasons. 'The official satellite positioning system of NATO is GPS. So British military forces are officially committed to using GPS. So what happens if another military system comes along? It gets very messy and it's been politically easier for European governments to steer away from it and say it's not a military system,' he says.

There is another reason why Galileo's military potential has been played down. It is written into ESA's charter that the agency will only work on projects that have non-military uses. Galileo is allowed because it is designed solely for civilian uses, even though it can clearly be used by the military too. 'Any new tool could be used in that way,' says Detain.

Some are concerned by China's 20% stake in Galileo, but

Detain says that only EU states will be given the secret codes to use the encrypted and highly accurate signals the satellites will broadcast. Such concerns may be academic, however. Galileo's publicly available signal will be nearly as accurate as the GPS military signals.

What really happened to Beagle 2?

The question is still open – but we now know that Britain's first spacecraft designed to land on another planet may have vanished into thin air, so to speak. *Beagle* was designed to sail into the Martian atmosphere on Christmas Day 2003 at 6 km a second, slow down with atmospheric drag, open first a pilot parachute and then the big one, then finally bounce to a standstill in a ball of air bags. *Beagle 2* was on autopilot. It was supposed to do things by the clock. So everything depended on scientists and engineers having timed the descent through the Martian atmosphere. Instead, there was silence.

One explanation is that the atmosphere of Mars may not have behaved in the way the models predicted. According to one instrument aboard the European orbiter *Mars Express*, the air density between 20 km and 30 km from the red planet's surface was a lot lower than predicted. But according to a different instrument aboard the NASA orbiter *Odyssey*, it was as predicted, which is why the jury is still out. But the thin-air hypothesis is also supported by the experience of the Americans, who landed two much heavier rovers, *Spirit* and *Opportunity*, successfully. Both landed down-range of their target zone, and both parachutes opened later than expected, which suggested that both made faster entries. 'What that is due to, the Americans aren't sure,' says Mark Sims of Leicester University. 'But it probably is a lower density in the atmosphere.'

If so, that explains why the *Beagle* team never heard from their baby. 'There is a whole nest of potential scenarios here. The pilot chute comes out too late and the main parachute comes out too late and we don't turn on the radar altimeter in time before we hit the surface. Any of these combinations are possible,' he says.

It isn't the only possible answer. Maybe space scientists did not really know how to calculate hypersonic and supersonic drag coefficients correctly, always a problem on a different atmosphere 100 million miles from home. If the *Beagle* team got another chance they might do things a bit differently: fit software that could react to the unexpected, for instance. 'There is a whole list of things we would change and it just depends on when the next opportunity is and how much time you have to change stuff,' he says. 'The bottom line is *Beagle 2*, on its parachutes, was to land at 16 m per second. If something went wrong high in the atmosphere you would land at 6 km a second, which is a bit different. At that speed there is very little you could do.'

Can you safely get a nuclear reactor into space?

Good question. NASA freely admits it hasn't got the foggiest. 'We don't know,' says Michael Braukus at the space agency's Washington HQ. 'It's premature right now to be talking about the impact, issues and things like that.' Which is interesting, as the agency is gearing up for possibly its most controversial launch yet: a spacecraft driven by a nuclear reactor.

NASA has awarded a $400 million contract to Northrop Grumman Space Technology to help design the nuclear-driven Prometheus project, which it hopes will blast off in a decade to explore the icy moons of Jupiter. Nuclear propulsion could send it further and faster with less fuel.

Chris Carr, a space technology researcher at Imperial College London, says that while existing space probes use radioactive power supplies, they are relatively safe because the material cannot trigger an explosive chain reaction if something goes wrong.

'The problem with a traditional nuclear reactor is that you could not build it strong enough to survive [an accident] because it's got to have delicate structures.'

Will we really find aliens by 2025?

According to Seth Shostak at the SETI (Search for Extra-Terrestrial Intelligence) Institute in California, we will. With the current pace of technology, it's just a matter of time. 'If you want to estimate when we're going to hear a signal, it only depends on two things,' he says. 'One, how many civilisations are there out there with the transmitters switched on, and secondly, how quickly are we doing our reconnaissance of the sky.'

Working out the first number is, unsurprisingly, the cause of much debate, with estimates ranging from zero to several million. Shostak goes for the more conservative numbers given by the Drake equation, a formulation of factors thought to be required for any life to exist. This suggests around 10,000 civilisations are advanced enough for us to find.

The second problem – how fast we are looking – is a technical challenge. 'At the moment, we're going very slowly,' says Shostak. 'We check out in the order of 50 to 60 star systems a year.'

'The bottom line comes out that even if you take Drake's more pessimistic [ideas], you'll trip across a civilisation by 2025,' he says.

What makes a planet a planet?

There are no hard and fast rules. But the 3,000 km-wide object spotted by American astronomers recently, officially called 2003 UB313 but nicknamed Xena by its discoverers, is a likely candidate for planet status. It is the biggest object found in the solar system since Neptune in 1846.

'There is no piece of paper which sets down what the definition of a planet is. Terminology is something you're left with from history and when you make new discoveries; it's hard to fit things into that terminology,' says astronomer Jacqueline Mitton.

A number of large objects have been found in recent years, including Quaoar (2002) and Sedna (2004) but neither of these is a planet. The original definition of a planet is a body that orbits the sun. How big it has to be is open to question – asteroids orbit the sun but no one would call them planets.

'One thing that some astronomers say definitely ought to define a planet is that it's got enough material in it that it naturally becomes a sphere,' says Mitton.

While the discovery of Xena is exciting, it casts doubt on the status of the current ninth planet, Pluto. Discovered in the early 20th century, it was given planet status because astronomers had no idea of its origins in a collection of asteroids called the Kuiper belt. Some astronomers argue that if this had been known at the time, it might not have been classed as a planet, but it's probably too late to change that. 'There's a feeling among a lot of the community that it would just be too confusing and upsetting to demote Pluto,' says Mitton.

The ultimate decision rests with the International Astronomy Union (IAU), which names heavenly bodies. But Mitton says popular will might overtake that decision: 'If the notion of this being the tenth planet catches on, then it probably will be the tenth planet, whatever the IAU says.'

What's the point of sending music into space?

In space, no one can hear you sing, yet human music has already travelled beyond Pluto to interstellar space aboard *Voyager 1* and *2*, launched in 1977. These each carried a gold record containing the sounds of Earth – surf, wind, thunder, whale calls, greetings in 55 languages – and music including Bach, Beethoven, Stravinsky, Chuck Berry's *Johnny B Goode* and *Dark was the night* by Blind Willie Johnson.

Had things worked out, humans might have heard Blur

blasting from Mars on Christmas day 2003. The band wrote the call sign for the ill-fated British lander, *Beagle 2*.

Guaranteed interplanetary smash hits headed towards the clouds of Saturn's mysterious moon, Titan, too. The European probe *Huygens* delivered, along with a package of sensitive instruments, specially commissioned tracks by Julien Civange and Louis Haéri. These are called *Lalala, Bald James Deans, Hot time* and *No love*. They were hits, if only because they slammed into Titan's clouds at 6 km a second. They went down well – three parachutes guaranteed them more than two hours of airtime. And they have already gone far – 3.2 billion km since their launch.

'The European Space Agency wanted to add artistic content to the mission, to leave some trace of humanity in the unknown and send a sign to any possible extraterrestrial populations,' says Civange, who has worked with the Rolling Stones, Simple Minds and David Bowie.

The tradition of bopping across interplanetary space is more than 30 years old. *Apollo* astronauts were 1960s explorers with 1960s tastes, and in 1972, the last *Apollo* crew woke up on their final morning on the moon to Richard Strauss's *Also sprach Zarathustra*.

What could you do to deflect an asteroid?

Not by sending up Bruce Willis in a space shuttle with a crew of veteran astronauts and oil drillers, that's for sure. The shuttle can only get to low Earth orbit and anyone who wants to deflect an oncoming asteroid will have to launch years ahead of any impact. So it's a task best left to an unmanned vehicle, built specially for the job.

Consider the problem: an asteroid on a collision course with the Earth could be detected 10 years ahead. Suppose it to have a diameter of 1 km. Suppose it to be travelling at a relatively sedate 39,000 kph. And suppose it to weigh about 1 bn tonnes. Anything that size hitting the Earth at an angle of, say, 45 degrees would, according to a University of Arizona website (lpl.arizona.edu/impacteffects/) generate the equivalent of a thermonuclear explosion of 50,000 megatonnes, enough to wipe out civilisation as we know it. Such collisions do occur, on average every half a million years. It would be a bad idea just to try to hit the thing with a nuclear warhead: even if the asteroid broke up, it would have time to reform and smash into the Earth anyway. But, experts have been pointing out for the last decade, provided the Earth had sufficient warning, this nemesis could be gently deflected. Last September, Imperial College London's asteroid expert Matt Genge calculated that something with the mass, acceleration and thrust of a Robin Reliant could push into a billion-tonne asteroid, with an acceleration of a billionth of a metre per second per second. If it did so for 75 days, it would change the asteroid's velocity by 0.7 cm per second, enough to make it miss its date with the Earth.

Can you see the Great Wall of China from space?

Not according to Yang Liwei, China's first man in space. 'The scenery was very beautiful,' Yang told Chinese TV when he returned to Earth. 'But I didn't see the Great Wall.'

His comments were taken, in some quarters, as proof that the story about the Great Wall (it being the only man-made structure that can be seen from space) is nothing more than a myth.

In fact, it is a myth, but not because the wall can't be seen from space. It is actually possible for astronauts a couple of hundred miles up in space to see lots of man-made structures, including skyscrapers, bridges and, weather permitting, the Great Wall.

All Liwei proved is that these objects can be difficult to spot. The Great Wall is especially tricky as it is a similar colour to the surrounding soil and is in a pretty bad state for large stretches. In places, it's difficult to see the Great Wall of China from China, even at fairly close range.

That's not to say it's impossible for astronauts to get a glimpse, and if Liwei wanted tips in finding it he could have asked Ed Lu, who lived on the international space station. 'It turns out you can see an awful lot from space,' Lu says on NASA's website. 'You can see the Great Wall. I've been trying, thus far unsuccessfully, to take a nice picture of it.'

Lu has had plenty of time to take in the sights from the space station windows. 'You can see the pyramids from space,' he says. 'With binoculars you can see an awful lot of things. You can see roads. You can see harbours. You can even see ships.'

Nobody at NASA knows where the idea about the Great Wall came from, but it was doing the rounds before the first satellite was launched. Another variation has it that the ancient stone border, just 6 m across, is the only artificial structure that can be seen from the moon. It can't. The few astronauts who have been there and looked back at the Earth report seeing a mass of white clouds and blue water, with patches of yellow sand and occasional flecks of vegetation.

Can stout shoes save you during a nuclear attack?

They might do, providing you shake the radioactive dust from them before going inside. The anti-fallout footwear is among several tips included in pamphlets passed on to a 1960s population convinced the world was heading for armageddon after the Cuban missile crisis.

The pamphlets suggest several ways of protecting yourself during a nuclear airstrike: stay in a sealed room, avoid going out and, if you do, wear a heavy coat and hat alongside your sufficiently stout shoes. And whitewash your windows to protect against 'nuclear flash': the moment when the bomb goes off and spits out intense amounts of heat and light.

Though it might all sound naïve, there is some merit in following these instructions if your neighbourhod happens to be at risk from a nuclear warhead. Nothing can survive a direct attack of course, but if you live several miles away from impact then how you react could save your life.

'Whitewashing and so forth does prevent a lot of thermal ignition of furniture and draperies,' says Paul Seyfried, president of Utah Shelter Systems and something of an expert – in theory at least – on keeping yourself safe during nuclear attack. 'It does go a long way to protecting you from the thermal fault and within about 6–8 miles, the thermal fault is going to cause a lot of bad burns to people who are outside.'

But whitewashing windows or sealing up rooms so that radioactive dust doesn't get in requires some time to prepare: with advanced intercontinental ballistic missiles, a target city would now be lucky to get more than the classic three-minute warning.

Getting worried? Seyfried designs and builds shelters just for these eventualities. A concrete bunker 8 ft underground would be more than enough to protect against a nuclear attack.

'The first three days is the real critical part. The rate of decay in the first two or three days is extremely rapid – if you could obtain effective shelter for at least three days, then the chances of avoiding radiation sickness are very good,' says Seyfried. 'After two weeks, the fallout radiation is only one percent of what it was an hour after the detonation.'

... and some other life & death questions

Can too much sleep kill you?

It seems so, if the results of a mammoth study following the lives and deaths of 100,000 people in Japan over the past decade are anything to go by. The Japanese researchers starkly state that people who snoozed for more or less than seven hours a night were more likely to die earlier – the third big project of its kind to make such a discovery. Exactly how this happens, though, is still unclear.

'When you fish out all the other variables it's very difficult to say whether it's the 9 hours or 5 hours sleep that is bumping you off,' says Jim Horne, director of the Sleep Research Centre at Loughborough University. Factors such as illness, stress and even unemployment can all influence both mortality and people's sleep patterns, although the new research did attempt to eliminate the effects of mental stress and depression. 'It just seems to happen that normal, healthy adults sleep for about 7 to 7½ hours a night,' Horne says.

The finding is the latest to trash the popular myth that

everyone needs at least 8 hours sleep. 'I've been in this game for 22 years and I still don't know where that came from,' says Neil Stanley, a sleep researcher at Surrey University.

Why do our submarines keep bumping into things?

The short answer is that it's hard to see where you're going under water, especially if you're trying to sneak around unnoticed, and accidents will happen. There is a fine tradition of minor prangs involving the Royal Navy's nuclear-powered fleet. *Trafalgar* was damaged after striking rocks off Skye in November 2002 and two trainee commanders steered *Triumph* into the seabed off Scotland in November 2000.

One problem is that submarines are not fitted with an automatic collision-warning device to use when submerged. Such a system would require keeping a sonar on at all times. Sonars emit those 'ping' sounds familiar from every submarine film ever made; if the pings bounce back, there is something in your way. If the something in your way is an enemy vessel, you'll have given away your position with potentially fatal consequences.

'There are conditions when you can have a quick transmit to see if there's anything in the way, but operational scenarios more often than not will not support that,' says Commander Jeff Tall, director of the Royal Navy Submarine Museum in Gosport. 'If you transmit on your own signal – if you go ping – some other bastard is going to hear it.'

Below periscope depth submarines effectively run blind.

They rely on 3D gyroscopes that sense acceleration and direction to compute position. This inertial navigation system needs to be calibrated regularly by surfacing for a satellite or visual fix.

'The most dangerous thing a submarine does is come up to the surface,' says Tall, who served for 30 years, commanding four submarines. Before surfacing, the crew prepare a 'tactical map' on which they plot objects that can be heard on the surface. Obstacles such as icebergs are difficult to hear. 'If he was in iceberg country [the sub] may well operate his high-frequency upward-looking sonar,' Tall says. This device sends out a pencil-thin signal in a specific direction.

Another trick is to raise the periscope while under water to look for shadows 'but if he's got glare from bright sunlight then he wouldn't see an iceberg and it only needs to be a few feet across to cause damage.'

Avoiding collisions in shallow water requires different skills. Submarines are fitted with bottom-sensing sonar but mostly rely on maps of coastal waters. They navigate using an imaginary 'pool of errors'. 'The longer you are without a proper fix then your pool of errors expands,' says Tall. 'You might get to the stage when you want to turn left, but your pool of errors touches a navigational hazard about 200 yards on the left. The only thing to do then is to come up and get a fix, or to take a chance.' Even the most careful commanders can get it wrong. Tall adds that as his submarine surfaced on one occasion, it appeared to have gone dark during the day. 'It was a supertanker sitting 100 yards away,' he admits.

What happened to Hunter S. Thompson's ashes?

Mercifully, most of Hunter S. Thompson probably landed near his home at Owl Farm, near Aspen, Colorado. It is very unlikely that he reached these shores.

The late author of *Fear And Loathing in Las Vegas*, in one last, spectacular gesture, had his ashes packed into a specially commissioned firework and fired from a cannon to 500 ft above the Rocky Mountains.

Humans and the molecules that compose them are recycled either slowly by burial or swiftly by cremation but even after death, their range remains limited. Those bits of them expelled into the atmosphere as carbon dioxide or other gases will continue to circulate. Mathematicians demonstrate as a lesson in probabilities that there is a 98.2% chance that your next breath will include an atom of air expelled by Julius Caesar when he took his last breath with the immortal words 'Et tu, Brute'.

But the more substantial fragments of the great exponent of so-called 'gonzo' journalism are more likely to have fallen straight back to Earth. His meteoric postmortem career was inevitably somewhat less dramatic than the career, for instance, of a meteorite. No one in Britain is likely to find a speck of charred Hunter as a smudge on the laundry.

'I've never exploded somebody's ashes,' says Matt Genge, a meteorite expert at Imperial College London. 'I doubt very much whether we'd actually receive any in this country. Meteorites tend to burn up at 40 km altitude, and you still find that the ablation debris are localised over several hundred kilometres.'

Is ball 38 luckier than the rest?

Don't put your house on it. A report in 2002 may have suggested the number 38 ball be 'physically examined' because it was sucked out of the machine so often, but the 100 or so draws since have seen it fall back into the statistically normal pack. Plus, if there were a physical anomaly with the ball, it would have to be repeated in each of the eight possible sets used for the main Lotto draw, of which one is selected at random each week.

Doubts over the random nature of the national lottery were raised by newspaper reports of an investigation by two statisticians at the University of Sussex. John Haigh and Charles Goldie, both members of the Royal Statistical Society, were asked by the National Lottery Commission to analyse the results of the first 637 draws since the game began in 1994.

In the subsequent report they noted that some combinations emerged with 'unusually high' frequency and that others showed 'major departures from randomness'. Most eye-catching, the number 38 had been drawn a surprising 107 times: the law of averages dictates each number should appear between 70 and 86 times in 637 draws.

'Unusual things do happen from time to time and this is one of them,' Haigh said. 'I suspect the National Lottery Commission hadn't thought things through. They hadn't considered what would happen if us wallahs were to tell them something fishy was going on.'

A spokeswoman for the National Lottery Commission says that research by the University of Salford's Centre for the

Study of Gambling showed that the 2002 result was a freak. 'All tests support the hypothesis of randomness. That is, they confirm no evidence of non-randomness among the Lotto draws,' the report concludes.

Miranda Pugh of lottery operator Camelot says the latex rubber balls were weighed and measured before use by, appropriately enough, the National Weights and Measures Laboratory in Teddington, Middlesex.

How much should they weigh? 'I can't give you details on that because it's secure information,' she says.

How many people does it take to be right?

This could get bumpy.

Does a consensus among a big group always point to what is right?

'It wouldn't make it right but it would make the opinion important in the sense that you've got all those people saying that,' says Paul Ernest, a philosopher at Exeter University. 'What would make it right would need to have a justification in terms of some accepted rules.'

In science, these accepted rules govern which new idea is accepted into the canon and which idea is thrown out as heresy. So consensus is important. But you only need to look at the example of Galileo (who proposed that the Earth orbited the sun and was imprisoned by the Church as a result) to see that consensus is not everything.

Christian List, a political scientist at the London School of Economics, distinguishes between two types of 'right'. In the first, the epistemic theory, there is an objectively 'right' answer to a question, based on moral facts that might or might not bear any relation to what people actually believe. In the second, the procedural account, there is no 'right' independent of what people believe. Democratic political systems are largely based on the latter argument.

'Ultimately, there is an issue of power involved,' says Ernest. He argues that, in science, if you can persuade the academics that you're right, then you're as good as right until someone else persuades them otherwise. Likewise, if Bob Geldof can persuade the G8 leaders that solving Africa's problems is the right thing to do, then, by them solving them, it becomes right.

Is Friday the 13th unlucky?

No, it is not specially unlucky. But for a substantial minority of Britons, the paraskevidekatriaphobia, or fear of Friday the 13th, regularly comes out of the closet.

'There has been no evidence that luck exists, in that no particular days are luckier than any other days,' says Richard Wiseman, professor of psychology at the University of Hertfordshire.

In 2003 he launched a survey on luck, which revealed that two out of three Britons thought themselves lucky or unlucky – and nine out of 10 subscribed to a superstition, such as touching wood. Those who felt unlucky tended to believe in superstitions about bad luck and were more anxious on Friday the 13th, and anxiety might just make accidents more probable.

'What matters is people's perception of it. Obviously, some people look at Friday the 13th and become concerned and anxious.'

The superstition almost certainly has its origins in Christian tradition. 'There are two things,' Wiseman says. 'One is the Last Supper, with 13 people at it. Also Christ was crucified on a Friday. Those two things are coming together.'

Do the dead contact us through tape recorders?

Lacking the basic requirement of proof, it's a safe bet to presume not, but that's no reason not to make a film about so-called electronic voice phenomena, or EVP, *White Noise* – a thriller in which a murdered woman leaves a message on a tape recorder identifying her killer.

A minority, not least Sarah Estep, president of the American Association of Electronic Voice Phenomena, believe such occurrences can be real. In the 1970s, Estep claimed to hear voices of spirits and aliens on tape.

EVP recordings are made either by leaving a tape recorder in a supposedly haunted room, or by detuning a radio so it picks up only noise. 'A lot of the time they're picking up voices of people going past,' says Chris French, a psychologist at Goldsmith's College, London. Other voices are merely radio interference, he adds.

According to French, there is often no speech, and people are imposing meaning on noise. 'We are exceptional at trying to find meaning in what we see and hear, but sometimes there is none. It's like when you're running the bath and you think you've heard someone call your name. You're just trying to find meaning when there isn't any.'

Can stress make your hair go grey overnight?

'I don't know the answer, but I do believe that people's observations are very difficult to discount – even if we can't explain them,' said David Fisher at the Dana-Farber Cancer Institute at Harvard Medical School.

What is clear is that genes have a role to play. Dr Fisher established in 1994 that greying is connected to the death of melanocyte stem cells that sit in hair follicles. These divide to produce pigmented cells called melanocytes which give the hair its colour. Once these have died there are no remaining stem cells to repopulate the hair follicles.

His team showed that two genes – Bcl2 and Mitf – are involved in this process in mice, and when they are mutated the mice go grey prematurely. It's not all about the genes though. 'Stress, drugs, life-changing experiences, may all activate alternate mechanisms,' said Dr Fisher.

Could brain implants control people remotely?

In theory yes, although you would not be able to do much useful with them. US researchers have produced a radio-controlled dogfish and plan to extend the technology to sharks, which raises the question of what the military, which funded the research, want to do it for.

The sharks, with their ability to glide quietly through the water and follow chemical trails, would make them ideal stealth spies. But could remote control be extended to people? The research, by Jelle Atema and his team at Boston University, reported in *New Scientist*, involved using a series of electrodes implanted in the brains of spiny dogfish to 'steer' the animals in a tank.

By stimulating the olfactory regions of the brain, the animals followed a 'phantom odour' created by the brain stimulation. A 2004 study from a different research group involved stimulating the nerves connected to rats' whiskers to achieve a similar effect.

So are remote-controlled human killing machines next on the list? No, said Richard Apps, a neurophysiologist at Bristol University. It's one thing to make a fish move left or right, but a huge leap to making someone perform a complex task against their will. 'That's the world of science fiction.'

Similar research has led to benefits for people though. Surgeons can now stop the debilitating tremors suffered by Parkinson's disease patients by implanting an electrode deep into their brains. The pattern of electrical signals from the device counteracts the neural signals to the limbs which cause the shakes.

And Matthew Nagle, a 25-year-old who was paralysed from the neck down in a knife attack, had a chip implanted in his brain that interprets electrical signals created by his thoughts and so allows him to control devices in the home such as his television.

These were only possible because of research on animals. 'You have got to understand the fundamentals first,' said Dr Apps.

How can you interrogate someone using the theme from Sesame Street?

You play it loud and you play it often.

The theme from the popular children's programme has reportedly joined heavy metal bands such as Metallica on the playlist of the US military, who are using loud music in an attempt to obtain information about possible weapons of mass destruction.

Playing loud music or white noise to prisoners is an established – if dubious – technique among interrogators. As well as disrupting sleep, the repetitive and strange sounds can distract and confuse captives through sensory overload, leaving them fed up and more willing to talk.

'We've got to have some peace and quiet in order to think, and with very loud noise of that sort it is harder to have normal thought processes,' says Dr James Thompson, a psychologist at the Middlesex Hospital in London, who has studied prisoners' reactions to torture. 'Most people will not be able to think logically in those circumstances.'

The type of music can also have an effect. 'Music that is discordant to the listener gives an additional element, but mostly it's got to be noisy, it's got to be repetitive and it's got to infuriate you,' Thompson says.

Many, including Amnesty International, argue that such methods should be classed as torture – and are therefore illegal.

British forces are effectively banned from using them since the former prime minister Edward Heath ordered an investigation into the use of loud music by British forces on 14 prisoners in Northern Ireland in the early 1970s.

Why has the army found it so difficult to replace bearskins?

Because hair is a complex thing and tough to reproduce artificially. But complaints from the animal rights lobby have forced the army to reconsider its affinity with bearskins, made from culled black bears in Canada, for 2,500 members of the Grenadier, Welsh, Irish, Scots and Coldstream Guards regiments.

For the past decade, the army has been looking at alternatives, to no avail. Fibres made from nylon either don't hold their shape, cannot withstand the weather, or lose their colour. Or they stand poker straight in the wrong electrical conditions.

Hair is notoriously hard to copy. 'There's a centre layer called the medulla and the area around that is the cortex,' says Barry Stevens, general secretary of the Trichological Society. 'You've got a whole series of cells around that which keep the whole thing together . . . The outer layer is rather like fish scales.'

In contrast, artificial hair – usually made of nylon – is just a smooth cylinder. It can't absorb water and the fibres don't stick together.

What is the most complicated music in the world?

According to scientists in Brazil and the US, it's not techno but the Javanese form of gamelan, followed closely by western classical music and the music of north India. A gamelan orchestra is based on metallic percussion instruments with wind and drums. In the Javanese version, musicians add a stringed instrument (called a rebab) and voices.

Heather Jennings, a physicist at Wesleyan University in Connecticut, led a team of researchers to develop a way to study the fluctuations in loudness in nine different genres of music.

Jennings has adapted a form of analysis normally used to monitor the patterns in heartbeats. After running each of the different styles of music through the analysis, she ended up with a number from zero to one that indicated its complexity – the closer the music came to one, the more complex it was.

Javanese gamelan and north Indian music got the highest values; Brazilian forro music and electronic techno music both came bottom of the complexity scale with values around 0.8, while jazz and rock 'n' roll sat roughly in between the two extremes.

According to Mark Lewis, a Southeast Asia author for the *Rough Guides*, gamelan is as interesting as it is complex. 'There's a mesmeric quality to gamelan, which stems from the repetitive nature of its rhythms,' he says. 'A single chime sounds familiar and ordinary, but many chimes combined create an effect that is other-worldly and rather beautiful. The overriding impression is of hearing running water.'

Jennings says the research is meant to introduce a new way to analyse music scientifically, rather than being a judgment of the music itself. 'None of the results has a direct bearing on harmony, melody or other aspects of music,' write the researchers. 'Our results apply only to loudness fluctuations, which can reflect aspects of the rhythm of the music.'

But not all complex rhythms are easy on the ear. On Jennings' scale, anything that has a value greater than one is essentially defined as noise.

Is it safe to fire a gun on a plane?

As long as the gun isn't pointed at you when it goes off, you should be OK. Worries that sky marshals might bring down a 747 by blowing a hole in the fuselage rather than a terrorist are largely unwarranted.

Industry insiders say marshals will be armed with guns that fire 'slow' bullets. The idea is that the bullets go fast enough to do serious damage to soft objects like humans, but slow enough so that, should they miss their target, they are less likely to rip a hole in the plane's fuselage. 'It's all to do with the construction of the bullet,' says Mike McBride, editor of *Jane's Police and Security Equipment*. Firstly, the bullet has a smaller firing charge, so it leaves the barrel much more slowly than a regular bullet.

Slow bullets differ from those used by the army in that they are not encased in a copper sleeve – a full metal jacket – that holds the bullet together, and they have hollow tips. This means slow bullets deform on impact, losing energy as they go, making them more likely to get lodged in their target rather than pass straight through, where they could hit something else.

Slow bullets are just one option open to sky marshals. Another bullet, called a frangible, is made of a ceramic compound that turns to dust if it hits something hard like the inside of an aircraft. 'They just fall apart, unless they hit something soft – and if that happens to be a human, they can certainly be lethal,' says McBride.

Even if regular bullets were used by sky marshals, the chances that a stray bullet would bring down a plane are minimal. 'People have got so het up about stray bullets passing through the shell of an aircraft, but people aren't going to be sucked out of a tiny 9 mm hole made by a bullet. And aircraft have got loads of holes in them already, it's part of their design. Another couple of holes aren't going to make much difference,' says McBride.

To cause any serious damage, the stray shot would have to take out the hydraulic system that is needed to fly the plane, something McBride calls 'a million-to-one shot'.

Sky marshals could make planes safer simply because the alternatives may be worse. 'If an aircraft is taken over by suicidal hijackers, then governments will be prepared to shoot it down rather than let it crash into a place full of loads of people,' says McBride. 'With that as the alternative, it's strange people get worried about having armed sky marshals on board.'

What can the way this is written tell you about the author?

Probably quite a lot, if you have enough text and sophisticated software. Evidence from a University College London study showed that Iris Murdoch was already showing signs of Alzheimer's disease when she wrote *Jackson's Dilemma*, her last novel. Clinical diagnosis came later, but computer-based comparisons of her earlier and final manuscripts showed that her vocabulary had reduced. Even before she knew it, Iris Murdoch was lost for words.

But textual analysis is already a well-established tool, and computer programs have begun to sort out a few outstanding literary mysteries. Chaucer composed his classic narrative poem long before the invention of printing, and scribes who made copies also made changes – accidentally or otherwise. The Canterbury Tales Project, based at De Montfort University, has been comparing manuscripts, in a patient attempt to establish which copies were oldest, and therefore closest to Chaucer's original words. Computer-aided analysis has been used to establish the authorship of Shakespeare's plays. But all the software has done so far is to eliminate Francis Bacon, Christopher Marlowe and Edward de Vere, Earl of Oxford.

Why is parting such sweet sorrow?

As Ella Fitzgerald sang, every time we say goodbye, we die a little. This is, of course, better than dying altogether, which may explain why there is sweetness even in sorrow. Juliet's lament about her separation from her very new acquaintance was enhanced by the thought that it would be temporary: that she would see Romeo again. In song and story, life goes on, even if something is missing.

Real life, however, offers no such certainties. In a here-today, gone-tomorrow world, there is a certain satisfaction in having existed at all. The exuberant joy of being is tempered by the wistful knowledge that nothing is forever. The Romans had a phrase for it: ave atque vale, hail and farewell. Psychological literature is rich in studies of separation anxiety, grief and loss. There is a denial phase to grief in which the bereaved believe that the lost one will come back. Then after a while, this denial gives way to anger, fear, guilt and depression. Loss of energy, fatigue, headaches and chest pains may occur before the inevitable adjustment.

Parting, neuropsychologists say, is a stretching of emotional bonds: the sorrow is tinged with the sweetness of the memories.

Index